Upcycling of plastic monomers by mixed microbial cultures

Von der Fakultät für Mathematik, Informatik und Naturwissenschaften der RWTH Aachen University zur Erlangung des akademischen Grades eines

Doktors der Ingenieurwissenschaften

genehmigte Dissertation

vorgelegt von

Master of Engineering

Romualdus Nugraha Catur Utomo

aus

Malang, Indonesien

Berichter: Univ. Prof. Dr.-Ing. Lars M. Blank

Prof. Dr. Nick Wierckx

Tag der mündlichen Prüfung: 12.05.2022

Bibliografische Information der Deutschen Nationalbibliothek
Die Deutsche Nationalbibliothek verzeichnet diese Publikation in der Deutschen Nationalbibliografie; detaillierte bibliografische Daten sind im Internet über https://portal.dnb.de abrufbar.

Romualdus Nugraha Catur Utomo:

Upcycling of plastic monomers by mixed microbial cultures

1. Auflage, 2022

Gedruckt auf holz- und säurefreiem Papier, 100% chlorfrei gebleicht.

Apprimus Verlag, Aachen, 2022
Wissenschaftsverlag des Instituts für Industriekommunikation und Fachmedien an der RWTH Aachen
Steinbachstr. 25, 52074 Aachen
Internet: www.apprimus-verlag.de, E-Mail: info@apprimus-verlag.de

Printed in Germany

ISBN 978-3-98555-088-3

D 82 (Diss. RWTH Aachen University, 2022)

Eidesstattliche Erklärung

Hiermit erkläre ich, Romualdus Nugraha Catur Utomo, dass diese Dissertation und die darin dargelegten Inhalte die eigenen sind und selbstständig, als Ergebnis der eigenen originären Forschung, generiert wurden. Hiermit erkläre ich an Eides statt

1. Diese Arbeit wurde vollständig oder größtenteils in der Phase als Doktorand dieser Fakultät und Universität angefertigt

2. Sofern irgendein Bestandteil dieser Dissertation zuvor für einen akademischen Abschluss oder eine andere Qualifikation an dieser oder einer anderen Institution verwendet wurde, wurde dies klar angezeigt

3. Wenn immer andere eigene oder Veröffentlichungen Dritter herangezogen wurden, wurden diese klar benannt

4. Wenn aus anderen eigenen -oder Veröffentlichungen Dritter zitiert wurde, wurde stets die Quelle hierfür angegeben. Diese Dissertation ist vollständig meine eigene Arbeit, mit der Ausnahme solcher Zitate

5. Alle wesentlichen Quellen von Unterstützung wurden benannt

6. Wenn immer ein Teil dieser Dissertation auf der Zusammenarbeit mit anderen basiert, wurde von mir klar gekennzeichnet, was von anderen und was von mir selbst erarbeitet wurde

7. Ein Teil oder Teile dieser Arbeit wurden zuvor veröffentlicht und zwar in ACS Sustainable Chemistry & Engineering 13 November 2020 (DOI: 10.1021/acssuschemeng.0c06019) und IOP Conference Series: Earth and Environmental Science 23 August 2021 (DOI: 10.1088/1755-1315/894/1/012024).

Aachen, 01.06.2022

Romualdus Nugraha Catur Utomo

Acknowledgments

My cordial gratitude goes to my supervisor Univ.-Prof. Dr.-Ing Lars M. Blank, for allowing me to work on this project. Thank you for your patient, kindness, support, and providing time to train me scientifically and personally. I am so grateful to have such a great experience working with you. I am also thankful to my second supervisor Prof. Dr. Nick Wierckx, for your support, guidance, and inspiring discussions. It is a wonderful experience to be part of the Institute of Applied Microbiology (iAMB). I thank *Lembaga Pengelola Dana Pendidikan* (LPDP) from Indonesian Government for the doctoral research scholarship.

I also would like to thank to iAMB members: Ulrike, Martin, Manja, Julia, Hendrik, Gisela, Isabel R., Annette, Till, amazing plastic group members (Birger, Kristina B., Gina, and Henric), Sandra, Wing-Jin, Ulf, Dieter, Tobias A., Salome, An, Caroline, Philipp D., Jana F, Simone, Mariam, Desy, Lisa, Kristina K. Carl, Tobias K., Conrad, Lena, Jonas, Max, Melanie, Nisha, Kalle, Vaishnavi, Juan, Benedikt, Birthe, Hao, Henrik, Vanessa, Bernd, Elena, Isabel B., Sebastian K., Andrea, Christoph L, and my students (Katrin and Benjamin). Thank you all for the sincere working relationship, scientific discussions, and individual assistance supporting this work.

My cordial appreciation goes to my family: my late father and mother, my beloved Mia, Hyosona, and elder brothers. Thank you for the prayers, support, and love that encourage me to finish this work. Most importantly, I thank God for giving me the healthy mind and body to keep going during this study.

Funding

lpdp

lembaga pengelola dana pendidikan

This work was funded by the *Lembaga Pengelola Dana Pendidikan* (LPDP) Scholarship, Ministry of Finance, Indonesia.

Table of Contents

Summary

Due to their robustness, versatile properties, and low price, plastics are used for such different applications as bone screws in medical applications and packaging in the food sector at ever-increasing amounts. The environmental problem caused by mismanaged plastic waste is prominent in many parts of this world, causing contaminations of the soil and marine ecosystems. To tackle this challenge, the end-of-life use of plastic must be considered in the value chain, including the fate of the polymers. A not yet industrially implemented possibility is to use waste plastic as a carbon source for microbes and produce value-added products. Ideally, the generated products have a higher value than the original plastic product and hence would not be just recycled but upcycled, a theme of the presented thesis.

Polyurethanes (PU) and polyethylene terephthalate (PET) are in the top 10 plastics produced and can either partially or fully be degraded using enzymes. These enzymes found in nature can hydrolyze these polymers, thereby releasing monomers like adipic acid, 1,4-butanediol, ethylene glycol, 2,4-toluenediamine, and terephthalic acid. The plastic monomers can be utilized as a carbon source by microorganisms like the biotechnological workhorse *Pseudomonas putida* KT2440 or new species like *Pseudomonas capeferrum* TDA1 to produce high-value products, like rhamnolipids, phenazine, and polyhydroxyalkanoates, presented in this thesis.

More specifically, this thesis presents a defined mixed culture utilizing the mock of a PU hydrolysate as a sole carbon source for growth and mono-rhamnolipid synthesis. The isocyanate derivative, 2,4-toluenediamine (TDA), exhibited strong inhibitory effects and hence was extracted prior to carbon use with 5 wt% di-(2-ethylhexyl) phosphoric acid (D2EHPA) in paraffin oil. The extraction of TDA was necessary, despite a member of the consortium could degrade this plastic monomer, indicating that the biochemical potential and the rates of the different members have to be taken into consideration in the design of mixed cultures.

P. capeferrum TDA1 can natively utilize adipic acid, 1,4-butanediol, and 2,4-toluenediamine, but not ethylene glycol and terephthalic acid. As an alternative approach, a consolidated mutant based on *P. capeferrum* TDA1 was established to utilize all monomers from PU and PET and to produce mono-rhamnolipids. Adaptive laboratory evolution was conducted to enable growth on ethylene glycol, while the *tph* operon from *P. umsongensis* GO16 was transformed to enable growth on terephthalic acid. The results are discussed in the context of the bow-tie structure of metabolism and plastic waste as a carbon source in a circular (bio)economy.

Zusammenfassung

Aufgrund ihrer Robustheit, vielseitigen Eigenschaften, und ihres günstigen Preises werden Kunststoffe für so unterschiedliche Anwendungen wie Knochenschrauben in der Medizintechnik und Verpackungen im Lebensmittelbereich in immer größeren Mengen verwendet. Das Umweltproblem, das durch unsachgemäß behandelten Plastikmüll verursacht wird, ist in vielen Teilen dieser Welt weit verbreitet und führt zu Kontaminationen des Bodens und der Meeresökosysteme. Um diese Herausforderung anzugehen, muss das End-of-Life von Kunststoff in der Wertschöpfungskette berücksichtigt werden. Eine noch nicht industriell umgesetzte Möglichkeit besteht darin, den Kunststoffabfall als Kohlenstoffquelle für Mikroben zu nutzen und Wertprodukte herzustellen. Im Idealfall sind die produzierten Produkte höherwertig als das Originalprodukt aus Kunststoff und werden daher nicht rezykliert sondern "upzykliert", ein Thema der hier vorgestellten Doktorarbeit.

Polyurethane (PU) und Polyethylenterephthalat (PET) gehören zu den Top 10 der hergestellten Kunststoffe und können teilweise oder vollständig enzymatisch abgebaut werden. Diese in der Natur vorkommenden Enzyme können diese Polymere hydrolysieren und dann die Monomere wie Adipinsäure, 1,4-Butandiol, Ethylenglycol, 2,4-Toluylendiamin und Terephthalsäure freisetzen. Die Kunstoffmonomere können von Mikroorganismen wie dem biotechnologischen Arbeitstier *Pseudomonas putid*a KT2440 oder neuen Spezies wie *Pseudomonas capeferrum* TDA1 als Kohlenstoffquelle genutzt werden, um hochwertige Produkte wie Rhamnolipide, Phenazine und Polyhydroxyalkanoate herzustellen, die in dieser Arbeit vorgestellt werden.

Diese Arbeit stellt eine definierte Mischkultur vor, die ein PU-Hydrolysat als einzige Kohlenstoffquelle für das Wachstum und die Mono-Rhamnolipid-Synthese verwendet. Das Isocyanat-Derivat 2,4-Toluylendiamin (TDA) zeigte starke Hemmwirkung und wurde daher vor der Verwendung von Kohlenstoff mit 5 Gew.-% Di-(2-ethylhexyl)-Phosphorsäure (D2EHPA) in Paraffinöl extrahiert. Die Extraktion von TDA war notwendig, obwohl ein Mitglied des Konsortiums dieses Kunststoffmonomer abbauen konnte, was darauf hinweist, dass das biochemische Potenzial und die Raten der verschiedenen Mitglieder bei der Gestaltung von Mischkulturen berücksichtigt werden müssen.

Als alternativer Ansatz wurde eine konsolidierte Mutante basierend auf *P. capeferrum* TDA1 etabliert, die alle Monomere aus PU und PET nutzt und Monorhamnolipide produziert. *P. capeferrum* TDA1 kann nativ Adipinsäure, 1,4-Butandiol und 2,4-Toluoldiamin verwenden, jedoch nicht Ethylenglycol und Terephthalsäure. Adaptive Laborevolution wurde durchgeführt, um das Wachstum auf Ethylenglykol zu ermöglichen, während das *tph*-Operon von *P. umsongensis* GO16 transformiert wurde, um das Wachstum auf Terephthalsäure zu ermöglichen. Die Ergebnisse werden im Kontext der Bow-Tie-Struktur des Stoffwechsels und Plastikmüll als Kohlenstoffquelle in einer zirkulären (Bio)Ökonomie diskutiert.

List of abbreviation

A. baylyi	*Acinetobacter baylyi*
Aa	amino acid
AA	adipic acid
ALE	adaptive laboratory evolution
a.u.	alternative unit
1,4-BDO	1,4-butanediol
2,3-BDO	2,3-butanediol
D2EHPA	di-(2-ethylhexyl) phosphoric acid
DMT	dimethyl terephthalate
EG	ethylene glycol
GC-FID	Gas chromatography-flame ionization detector
HDPE	High-density polyethylene
HPLC	High-performance liquid chromatography
iAMB	Institute of Applied Microbiology
IGV	Integrative genomics reviewer
InDel	Insertion deletion polymorphism
LDPE	Low-density polyethylene
LPDP	Lembaga Pengelola Dana Pendidikan
Mcl	medium chain length
MDI	4,4-methylene diphenyl isocyanate
MHET	mono(2-hydroxyethyl) terephthalic acid
MSM	mineral salt medium
OD_{600}	optical density 600
PCA	phenazine-1-carboxylic acid
PE	polyethylene
PET	polyethylene terephthalate
PHA	polyhydroxyalkanoates
PHZ	phenazine
PP	polypropylene

PS	polystyrene
PU	polyurethane
PVC	polyvinylchloride
PYO	pyocyanin
RWTH	Rheinisch-Westfälische Technische Hochschule
SNP	single nucleotide polymorphism
TA	terephthalic acid
2,4-TDA	2,4-toluenediamine
TDI	2,4-toluene diisocyanate
ULBs	Urban local bodies
wt%	weight percent

List of figures

List of tables

Chapter 1

Introduction

Contributions:

This chapter was written by Romualdus Nugraha Catur Utomo and reviewed by Lars M. Blank.

1.1 Global plastic production

Over the decades, global plastic production has been exponentially increasing, and the increased demand is due to its durability, widespread applications, low price, and societal benefits (Andrady and Neal, 2009; Plastics Europe, 2020). Based on the latest data, the global production of plastic has been reaching 368 million metric tons in 2019, and within a year, has been increasing by another 9 million metric tons (Plastics Europe, 2019, 2020). In contrast, plastic production decreased in Europe from 61.8 million metric tons in 2018 to 57.9 million metric tons in 2019. The decrease could be caused by many factors, such as a possible effect of the increased plastic export from the US. On the other hand, the consumption increased by 0.4%, mainly coming from the packaging and construction sectors. However, it is still premature to deduce a trend in the long term, for example, aimed for by policies against single-use plastics (Plastics Europe, 2020).

During the COVID-19 pandemic, plastic production in Europe also significantly declined in the first half of 2020. However, the production went up in the second half of 2020 and is predicted to increase slowly in 2021. Nevertheless, it depends on the recovery of the many industries from the pandemic and depends on the demand from primary customers, such as the automotive, building, and construction sectors (Plastics Europe, 2020). As a unique polymer, polyurethane continue to be manufactured, with high demand in the building, construction, and automotive sectors. Meanwhile, leading polymers, polyolefins (polyethylene (PE) and polypropylene (PP)), together with PET and polystyrene (PS), give a significant contribution to the packaging sector (Plastics Europe, 2020).

1.1.1 Polymer resins

Resin, which is made from crude oil, is the core ingredient for almost all plastic products. Furthermore, compounds or monomers are placed under heavy pressure and mixed with catalyst and additives until it forms liquid and eventually becomes plastic pellets. PU, PET, high-density PE (HDPE), low-density PE (LDPE), linear low-density PE (LLDPE), PP, PS, and polyvinylchloride (PVC) are categorized as polymer resin-based plastics. PET, PE, and PP have been used in packaging, an application contributing 42% of non-fiber plastics (Geyer *et al.*, 2017; Plastics Europe, 2020).

Table 1 Synthetic polymer categorization from resin-based plastic

Type of polymer	Reversibility	Recyclability	Contribution in Global plastic production (%)	Main material/ monomers for synthesis
PU	X	X	7.9	*Adipic acid, ethylene glycol, 1,4-butanediol, 2,3-butanediol, 2,4- toluene diisocyanate, 4,4-methylene diphenyl isocyanate*
PET	√	√	7.9	*Ethylene glycol, terephthalic acid*
HDPE	√	√	12.4	*Ethylene*
LLDPE, LDPE	√	√	17.4	*Ethylene*
PP	√	√	19.4	*Propylene*
PS	√	√	6.2	*Styrene*
PVC	√	√	10	*Vinyl chloride*

The family of polymer resins can be categorized into two groups based on their reversibility after heating, namely thermoplastic or thermoset. Thermoplastic is a reversible plastic that can be melted when heated and hardened when cooled, such as PET, HDPE, LDPE, PP, PS, and PVC. In application, it can be reheated, reshaped, and frozen repeatedly. Whereas thermoset is a family of plastics that always remain in the solid state, it cannot be remelted as well as remolded, such as PU, melamine resins, and epoxy resins. Meanwhile, thermoplastic polyurethane (TPU) also exist, which are different from the usual elastomer due to the absence of crosslinks (Huntsman, 2010; Plastics Europe, 2020).

The challenges of recycling of PU comes from the varying monomer mixtures, and ever-changing additives, while the reacted isocyanates and their derivatives namely 4,4-methylenediamine and 2,4-toluenediamine are a challenge in them self. Nevertheless, polymer degradation processes via solvolysis and pyrolysis are reported (Xanthos and Patel, 1998; Garrido and Font, 2015).

Previously, polystyrene (PS) is considered non-recyclable plastic due to its heat resistance, but now a solvent-based purification process has been well developed and scaled. The technology, the so-called Creasolv Process, starts with PS-foam demolition waste, which is dissolved in solvents. The solid impurities are then separated through filtration and subsequently incinerated. The next step is crucial, allowing separation from the additive hexabromocyclododecane (HBCD). Once the polymer is purified, it is transferred into the granulated polymer and the new additives, distilled, and reused in a closed loop. At the same time, the HBCD remains in the sludge and is treated in the Bromine Recovery Unit (BRU) (Polystyreneloop, 2021).

A significant problem of PVC recycling mostly comes from the high chlorine content and high level of hazardous additives in this material. Since more PVC products are reaching their end-of-life, the recycling process of PVC is urgently needed. Currently, this material is considered a recyclable material, and there are two possible ways to recycle, namely mechanical and feedstock recycling. The first recycling involves mechanically treating waste by reducing it into smaller parts. The resulting granules, called recyclates, can be melted and remolded into different products. The second recycling involves chemical processes, such as pyrolysis, solvolysis, and heating, which could convert the waste into its chemical components. Eventually, possible products such as sodium chloride, calcium chloride, and hydrocarbon products can be generated and used for new PVC manufacture (Vinylplus, 2014).

1.1.2 Synthetic fibers

Together with polyester and polyolefin fibers, polyamide and polyacrylic are prominent examples of fiber-based plastics with 10% of the total global plastic production. The fiber plastics are most likely recyclable. Ester, amide, and acrylic are the most common primary materials for the synthesis (Geyer *et al.*, 2017). Additives, which mostly come from plasticizers, fillers, and heat stabilizers, and other polymers, had contributed about 8.8% of global plastic waste in 2015 (Jambeck *et al.*, 2015; Geyer *et al.*, 2017; Law *et al.*, 2020).

1.2 Global plastic waste in landfills and natural environments

From the data between 1955 and 2050, the resin plastic waste is predicted to increase exponentially to reach over 25,000 million metric tons with 6,000 million metric tons waste of fiber plastics and 2,000 million metric tons waste of additives at the end of the period. From that amount, 9,000, 12,000, and 12,000 million metric tons would be recycled, incinerated, and just discarded into landfills or natural environments, respectively (Geyer *et al.*, 2017).

Plastic waste today is a severe worldwide problem, as its quantity significantly increases per year, while the quality of end-of-life treatment differs significantly among countries. Plastic waste distribution is also related to the population of certain countries and the production itself. As we know, more than half of plastic is manufactured in Asia, followed by North America and Europe with 19% and 16%, respectively (see figure 1). The plastic waste distribution is also caused by the trade of the waste itself with Asian countries as a targeted destination (Jambeck *et al.*, 2015; Plastics Europe, 2020).

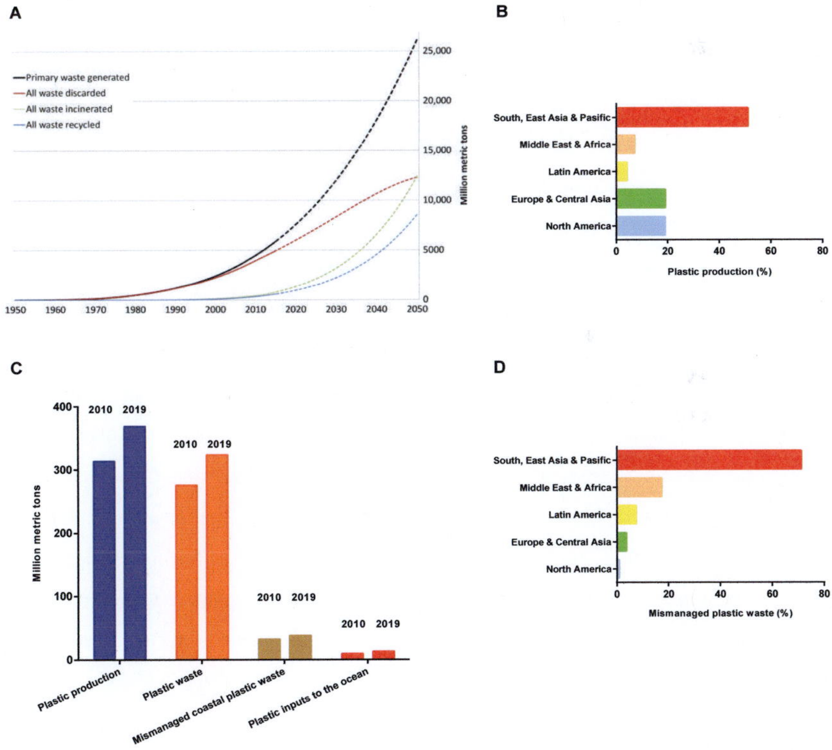

Figure 1 A: Accumulation of plastic waste between 1950 and 2050, **B:** Distribution of global plastic production in 2019 (in percent), **C:** Estimated mismanaged plastic waste per year in

the last decade (in percent), and **D:** The change of global plastic production and its waste in the last decade (in million metric tons) (Plastics Europe, 2020).

1.2.1 Dominant contribution of plastic waste generation by Asian countries

The management of plastic waste, including incineration and recycling, is recorded better in Europe and the US than in Asia. In 2014, Europe reportedly had an incineration and recycling rate of around 40% and 30%, respectively. China, the leading country in plastic production, had incineration and recycling rates of about 30% and 25%, respectively, in the same period (Geyer *et al.*, 2017). Due to limited data, it is only reported that open dumping and open burning are common in Southeast Asian countries. On the other side, recycling of plastic waste is only conducted by the informal sector (Jain, 2017; Modak *et al.*, 2017).

The significant amount of plastic production, plastic waste generation, and inefficient plastic waste management, including recycling by most Asian countries, contributed to the amount of worldwide mismanaged plastic waste. The amount of plastic waste that directly contaminated oceans was estimated at 4.8 to 12.7 million MT in 2010. From that amount, 6,350-245,000 MT are floating plastic waste (Jambeck *et al.*, 2015; Plastics Europe, 2020). Among the top 10 countries contributing directly to plastic waste into the ocean, China is leading with 8.82 million MT and Indonesia with 3.32 million MT. However, Indonesia would face a more serious problem because its coastline is longer than that of China, 54,720 km compared to 14,500 km (Jambeck *et al.*, 2015).

As the second most plastic waste contributor, Indonesia would accept more global critics regarding ocean contamination. That would directly and significantly impact China's implementation of a new policy banning plastic waste import. Beforehand, China was notified as a country capable of importing plastic waste to be recycled as newly manufactured products such as clothing and toys. Consequently, Indonesia with Thailand, Malaysia, and Vietnam, would receive more than 5% of global plastic waste previously exported to China. In contrast, the proper management of plastic waste in those countries is still low (Brooks *et al.*, 2018).

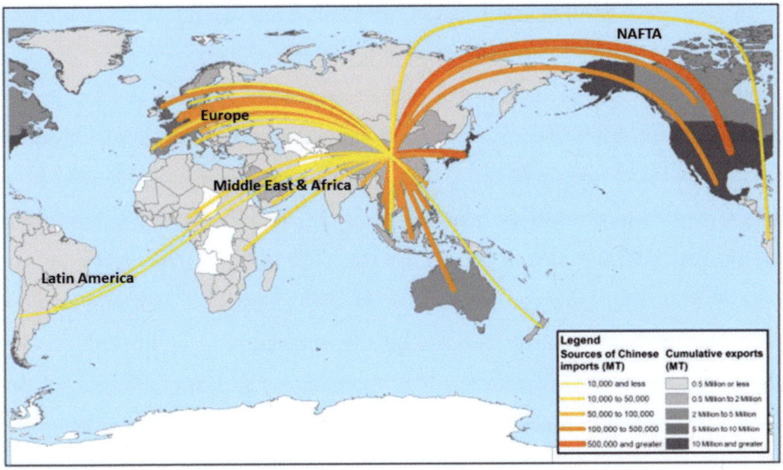

Figure 2 Source of plastic waste import to China in 2016 and cumulative plastic waste export between 1988 and 2016 (in million MT) (Brooks *et al.*, 2018).

1.2.2 Plastic waste in Indonesia and its domino effects

As an archipelago country, Indonesia has the most significant contribution of plastic waste among Southeast Asian Nations (ASEAN) to the marine environment, which corresponds to the condition of some big rivers in Indonesia. It was recorded that four-Javanese rivers in Indonesia stood on the top 20 - most polluted rivers in the world, namely, Brantas (7th), Solo (10th), Serayu (14th), and Progo (19th) with annual plastic waste emission as much 38,900, 32,500, 17,100, and 12,800 tones, respectively. This amount of plastic waste would significantly contribute to the global mismanaged plastic waste. This condition of the plastic industry has been putting Indonesia in a predicament (Lebreton *et al.*, 2017).

In 2016, the Indonesian government tried to reduce plastic use by adding a tax of 2 US cents on a single-use plastic bag. However, it could not significantly reduce plastic use in society and the plastic waste in the environment. At the same time, there were increasing critics regarding the transparency in how the revenue was used. The unclear hierarchy of plastic waste regulation from the central government also affects local governments' plastic use and plastic waste control. Regarding high contamination of plastic waste and a lack of plastic management, the recycling rate was recorded as low as 7% in 2015. The remaining wastes were disposed (69% landfilled, 5% burned, 10% buried) (Modak *et al.*, 2017).

As a result of mismanaged plastic waste in Indonesia, there were a lot of negative impacts. The plastic waste in the sea has generated losses of as much as $ 2.1 billion in the fisheries, shipping and tourism sectors, and insurance business (Ritchie and Roser, 2018). The fishery's loss came from intestinal injury, death, and toxic accumulation in the food chain caused by plastic waste. The floating waste also can inhibit the mobilization of fish and reduce the intensity of sunlight and oxygen, which are essential in respiration. Generally, plastic waste has also caused some damage to boats by the entanglement of the rubbish to ship components, such as plastic waste ingested into intakes for cooling engines and causing propeller damage. As a consequence, there are many fishing vessel claims to the insurance companies due to accidents caused by marine debris, especially plastic waste (McIlgorm *et al.*, 2009). The negative impact of plastic waste in the tourism sector is caused by the low aesthetic value of coastline or beaches for visitors and residents. This condition has been happening in Bali, as the primary destination of overseas tourists to Indonesia (McIlgorm *et al.*, 2009).

1.3 The deadly effects of global plastic contamination on life

As a result of global plastic contamination in the ocean, the waste reportedly had adverse effects on organisms' lives, especially eukaryotic. Two hundred sixty-seven species of fish, sea turtles, sea birds, and marine mammals are affected by the plastic waste in the ocean (Clean Water Action, 2021). Deaths are mainly caused by ingestion, infection, suffocation, and entanglement of plastic (Taylor-Smith, 2018). Tragically, fish in the north pacific reportedly has been ingesting 12,000 to 24,000 tons of plastic per year (The Center for Biological Diversity, 2018). The invertebrates, such as commercial crustaceans and Echinodermata are also known to ingest microplastics. Microplastics were found in the stomach, hepatopancreas, ovary, and gills of green crabs (*Carcinus maenas*) and sea urchin. The deep sea creatures also suffered from plastic waste, mostly from polyethylene and polyester from shopping bags and clothes in microplastic form (Taylor-Smith, 2018).

From stranded wild-sea-turtles, macro- and microplastics were found in extracted feces (Ostiategui-Francia *et al.*, 2016). A California grey whale has been reported dead in 2010 after autopsies found more than 20 plastic bags, small towels, a pair of pants, a golf ball, duct tape, and surgical gloves inside the stomach (Clean Water Action, 2021). The observation of dead Laysan Albatross chicks between 1994 and 1995 on Sand Island, Midway Atoll showed that

97.6 % of 251 dead chicks contained plastic in their stomach and/or intestines (Auman *et al.*, 1997).

1.4 Plastic waste recycling

The recycling of polymers can be categorized as primary (mechanical), secondary (mechanical), tertiary (feedstock or chemical), quaternary (incineration), biological recycling, and combination of them.

1.4.1 Primary mechanical recycling

The principle of primary mechanical recycling is the direct reuse of discarded and uncontaminated plastic products to produce new products without changing the structure and loss of properties. Some steps are required in advance, such as grinding, shredding, crushing, and milling. Mainly, the primary mechanical recycling is conducted by the manufacturers of plastic products. After the melting step, the recyclate can be obtained from this recycling and remolded to new products. Therefore, this recycling can only be applied on thermoplastic polymers, such as PET, PE, PP, and PVC (Ignatyev *et al.*, 2014).

Several studies have shown that plastic waste can be added to asphalt composition. The plastic waste is used as a filler of asphalt concrete mixture, while PP, LDPE, HDPE, and PET are predominantly used in those studies. One reported study showed the possibility of using plastic waste as filler to substitute the Portland cement filler in submersed conditions in freshwater and seawater. The use of plastic waste in infrastructure construction such as road and laminated roofing would significantly help to consume a large amount of plastic waste (Gawande *et al.*, 2012; Maal *et al.*, 2017), while potential release overt time has to be considered.

1.4.2 Secondary mechanical recycling

From mixed plastic waste, exact content and purity are hardly known; therefore, secondary mechanical recycling is needed, which involves more steps, such as separation and purification. Due to chain scissions, the polymer's molecular weight reduces and may lead to less favorable mechanical properties during this process. As in primary mechanical recycling, only thermoplastic can be reprocessed.

1.4.3 Tertiary or feedstock recycling

Tertiary or feedstock recycling can be defined as a recycling process that converts the polymer chains to smaller molecules via chemical processes. Solvolysis, pyrolysis, and gasification are well-known tertiary recycling methods, which have widespread applications that are not limited to thermoplastic.

Based on the solvent used, solvolysis can be categorized as alcoholysis, hydrolysis, acidolysis, aminolysis, and various chemical reactions to depolymerize the polymers to oligomers or monomers (Xanthos and Patel, 1998). Those processes are mostly applicable for polymers, such as polyester, polyamide, and polyurethane.

Pyrolysis and gasification are currently conducted to depolymerize plastic waste in reactors without requiring much oxygen and more environmentally friendly than open combustion or incineration. Both are thermal conversion methods that are intended to minimize the pretreatment steps of plastic waste. In pyrolysis, the plastic waste is thermally decomposed to combustible gases and varied oils depending on the type of polymer and operation environment of pressure, temperature, and heating rate. On the other side, gasification is the process of obtaining syngas (Ignatyev *et al.*, 2014; Garrido and Font, 2015; Shakorfow, 2016). Both have pollutants in reduced form, namely COS and H_2S, while incineration or combustion has pollutants in oxidized form (World Energy Council, 2016).

1.4.4 Quaternary recycling or incineration

Incineration (combustion in a dedicated plant) needs sufficient oxygen to thermally convert the plastic waste to a flue gas at high temperatures. This process is mainly applied for mixed and heavily contaminated waste, which cannot efficiently and economically be recycled (Shakorfow, 2016). In principle, burning the dense waste can produce heat, electricity, or other forms of energy. As an example of the current application, the heat produced during incineration can generate steam for driving the turbine to produce electricity. Environmental assessment has been strictly conducted to minimize the pollutant towards the surroundings (Umweltbundesamt, 2008). Indeed, the most modern incineration plants are located centrally in cities, to minimize logistic efforts and to maximize energy use, e.g., by heating local houses.

1.4.5 Biological degradation and upcycling

The plastic waste can be degraded in the environment through four mechanisms, namely photo-degradation, thermos-oxidative degradation, hydrolytic degradation, and biodegradation by microorganisms (Webb *et al.*, 2012). In principle, plastic polymers are very recalcitrant to be degraded directly by microorganisms. Therefore, to increase the biodegradation rate, the bioavailability of plastics can be optimized by the action of abiotic factors, such as low salinity and high temperature. Those optimum abiotic factors can be easily reached in freshwater ecosystems and tropical regions. Low crystallinity and high hydrophilicity of a polymer also significantly improves enzymatic degradation (Wierckx *et al.*, 2018; Wang *et al.*, 2021).

The biodegradation of plastics consists of several steps: bio-deterioration, bio-fragmentation, assimilation, and mineralization (Dussud and Ghiglione, 2014). Based on reported results, the degradation of PS, PE, and PP were below 15% weight, while there was no successful biodegradation of PVC. Those reported results came from the biodegradation by using *Exiguobacterium* sp for polystyrene, *Brevibacillus borstelensis, Rhodococcus ruber* C20, *Arthrobacter* sp., *Pseudomonas* sp., *Bacillus* sp. for polyethylene, while *Bacillus* sp. and *Pseudomonas* sp. are prominent species for the biodegradation of polypropylene (Wierckx *et al.*, 2018). All the reports on biodegradation of C-C bonds containing plastics have to be taken with care, as the reducing power of enzymes is too low for this catalytic achievement (Jiang *et al.*, 2021), but rather analytical artifacts and the degradation of oligos and additives are reported.

Polyurethane (PU) is a complex material manufactured from various monomers and is recalcitrant. Mainly, they consist of long polyols, isocyanate, and chain extenders; therefore, biodegradation depends on the monomers' chemical bonds and properties. While filamentous fungi, *Aspergillus tubingensis,* and *Aspergillus* sp. strain S45 (Khan *et al.*, 2017; Osman *et al.*, 2018) were reported to be able to degrade polyester PU film in the *ex-situ* condition, some *Cladosporium* species, *Aspergillus fumigatus*, and *Penicillium chrysogenum* were known to be able to degrade polyether PU foam (Álvarez-Barragán *et al.*, 2016). However, we should consider that no etherase was reported so far for PU degradation.

Plastic waste upcycling is considered one of the best strategies to significantly reduce plastic waste and keep the carbon in the envisaged circular bioeconomy. The circular bioeconomy would be a long-term strategy for accelerating and separating economic growth from the environmental biodegradation of polymers (Hetemäki *et al.*, 2017; A. Glaser, 2019). The upcycling process, unlike the current plastic recycling process, aims to generate higher value products from plastic waste, which allows additional efforts to use, e.g., less sorted plastic fractions. Upcycling is reported using chemo- or biocatalysis. The enzymatic depolymerization of polymers results in oligomers or monomers. These monomers can either be purified for the next round of polymerization or be used as sole carbon sources to convert them into valuable products, e.g., biodetergents using microbes as catalysts. Utilizing plastic waste monomers as a sole carbon source for bioplastic production, such as polyhydroxyalkanoates (PHA) (Kenny *et al.*, 2008, 2012; Franden *et al.*, 2018; Li *et al.*, 2020; Narancic *et al.*, 2021) is a promising strategy in a circular plastic economy.

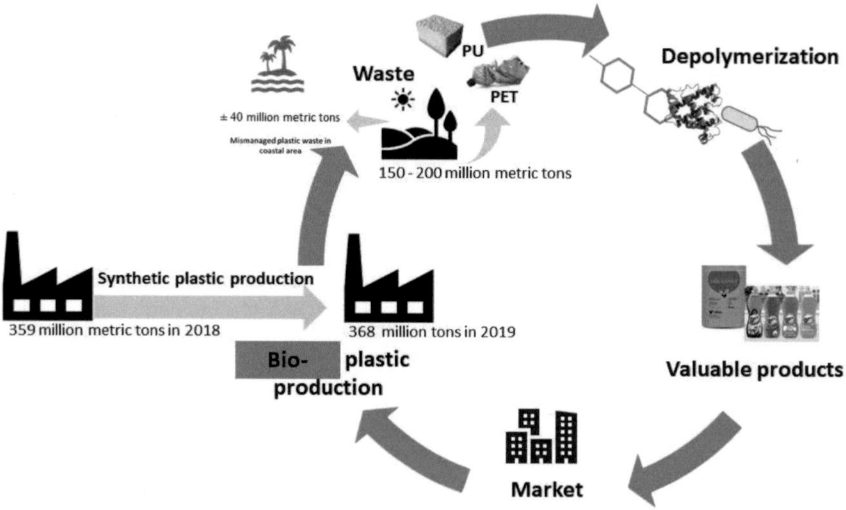

Figure 3 The route of plastic waste upcycling. The circular bio-economy route consists of five steps, namely (synthetic and or bio-) plastics production, plastic waste generation, plastic waste depolymerization, utilization of plastic waste monomers as substrates for generating valuable products, and market value of the valuable products.

1.5 Polyurethane upcycling

PU is widely used in the medical, automotive, and industrial sectors (Nikje *et al.*, 2011) In 2015, the PU's global production reached 27 million metric tons, making it the 6th largest polymer group. Many PUs are highly stable, with long-term applications in the housing sector, which is apparent in the fact that only half of the produced PU has been generated as waste

(Geyer *et al.*, 2017). However, end-of-life management options for PU waste are currently limited.

1.5.1 Structure of PU: polyols, isocyanate, chain extenders, and additives.

PU is composed of soft segments consisting of long polyester or polyether-based polyols and hard segments comprising the isocyanate groups and the chain extenders (Magnin, Pollet, *et al.*, 2019). Polyols increase the flexibility of PU, while isocyanates and chain extenders give the rigidity to the chain (Szycher, 2012). In addition, the polyols have a significant effect on the hydrolytic stability and mechanical properties of PU (Sonnenschein, 2014). On the other side, isocyanates play a vital role in PU manufacturing, with methylene diphenyl diisocyanate (MDI) and toluene diisocyanate (TDI) as most frequently used monomers because of their high reactivity (Gama *et al.*, 2018). Several PU additives are also required for manufacturing and end product performance, such as blowing agents, preservatives, catalysts, and flame retardants (Gama *et al.*, 2018).

1.5.2 PU depolymerization

Solvolysis with, e.g., glycol or methanol, is investigated extensively to depolymerize PU waste, including rigid and flexible foams (Zhang *et al.*, 2018). The process mainly consists of four steps: grinding, glycol addition (depolymerization), alkoxylation, and polyol recovery (Sheel and Pant, 2018). As alternatives, pyrolysis and hydrolysis are reported for the chemical recycling of PU. Pyrolysis is conducted in oxygen-free furnaces at elevated temperatures (600°C - 900°C) (Garrido and Font, 2015). Hydrolysis uses water (150°C - 200°C) or steam (200°C - 320°C) for depolymerization (Sheel and Pant, 2018). Pyrolysis generates mainly gaseous products such as methane and ethylene, while alcoholysis, ammonolysis, and hydrolysis generate free polyols, diamines, and carbon dioxide as end products (Garrido and Font, 2015; Sheel and Pant, 2018; Zhang *et al.*, 2018).

PU is a heterogeneous class of polymers regarding monomer composition, molar ratios, and additives used. This heterogeneity allows PU manufacturers to tailor the polymer to the individual applications and the customers' requirements (Gama *et al.*, 2018). Common monomers are adipic acid, 1,4-butanediol, ethylene glycol, and the isocyanates 2,4-toluene diisocyanate (TDI) or 4,4-methylene diphenyl isocyanate (MDI). TDI is mainly used to manufacture flexible foams, such as upholstery, mattresses, and vehicle seats (PlasticsEurope,

2012). Upon (enzymatic) depolymerization of PU, the isocyanate monomers are released as the corresponding amine (Magnin *et al.*, 2020). The resulting 2,4-toluenediamine is toxic to humans and was classified as a carcinogen in animal studies. The compound is also water-soluble (38 mg/L at 25°C) and, therefore, can easily contaminate soil or other environments (European Union Risk Assessment Report, 2008).

1.5.3 Bioconversion of PU monomers

Nowadays, PU's enzymatic depolymerization is receiving increasing attention as a green solution, without the need for organic solvents and chemical catalysts (Cregut *et al.*, 2013; Magnin *et al.*, 2020). Although still in its infancy compared to chemical recycling, especially for thermoset and polyether PU, enzymatic depolymerization could enable a completely biological upcycling approach. The approach is a promising end-of-life solution for PU, considering the limited mechanical recycling options for this polymer class. A mixture of four hydrolases, namely LC cutinase (LCC), TfCut2, Tcur1278, and Tcur0390, could foster the complete degradation of the ester bond of polyester PU (Schmidt *et al.*, 2017). However, polyether PU is quite recalcitrant towards enzymatic depolymerization, even though urethane bond cleavage in this plastic was recently reported (Wei and Zimmermann, 2017b). Combining esterase and amidase activity improved the hydrolysis of the urethane bond further (Magnin, Pollet, *et al.*, 2019). A *Penicillium* was reported as the best degrader of thermoplastic polyurethane (TPU) and had the highest esterase activity on the substrate with six aliphatic carbons (Magnin, Hoornaert, *et al.*, 2019). However, an industrial enzymatic degradation process of PU is still elusive and requires considerable progress in enzyme technology. The selective degradation of chemical bonds in PU might allow oligo recycling, which would contribute to the vision of isocyanate free PU synthesis (Magnin, Entzmann, Bazin, *et al.*, 2021; Magnin, Entzmann, Pollet, *et al.*, 2021).

Microbes from the genus *Pseudomonas* are promising biodegradation agents for synthetic plastic waste monomers (Wierckx *et al.*, 2015; Wilkes and Aristilde, 2017). *P. putida* KT2440 strains that utilize ethylene glycol and 1,4-butanediol have been engineered and or evolved (Franden *et al.*, 2018; Li *et al.*, 2019, 2020). In a recent study, a novel strain, *P. capeferrum* TDA1 was reported utilizing 2,4-toluenediamine, a common precursor of TDI for polyurethane synthesis (Cárdenas Espinosa *et al.*, 2020). In addition to *Pseudomonas, Acinetobacter baylyi* ADP1 is a prominent model in molecular biology for the last 25 years, and this strain is natively able to utilize adipic acid, a common monomer for PU polyols synthesis (Parke *et al.*, 2001).

1.5.4 Application of microbial mixed culture in biotechnology

The microbial degradation of a monomer mixture (or plastic hydrolysate) is not simple, as individual microbes often lack the required biochemical pathways, or these pathways are not efficiently expressed. This can be due to the misregulation of the non-natural carbon source (Li *et al.*, 2019, 2020), or due to carbon catabolite repression (Vinuselvi *et al.*, 2012). The latter often dictates which carbon source is preferentially used under a certain growth condition, reducing co-utilization as required in biotechnology to a minimum. Therefore, mixed microbial cultures are discussed to increase the biochemical space and overcome the limitations of individual microbes (Wang *et al.*, 2019). The application of mixed microbial cultures in biotechnology is motivated by biogas and waste treatment plants, fulfilling biochemical transformations on ever-changing substrates (Kleerebezem and van Loosdrecht, 2007; Agler *et al.*, 2011). Applications of microbial mixed cultures in academia are numerous, like biosurfactants production and biodegradation of 4-nitroaniline from textile dye wastewater (Khalid *et al.*, 2009; Antoniou *et al.*, 2015), but they have to prove their competitiveness against alternative approaches. In industry, the two organism consortium of *Ketogulonicigenium vulgare* and a *Bacillus* species for the synthesis of the vitamin C intermediate 2-keto-L-gulonic acid from L-sorbose is still a rare but prominent example for defined microbial mixed culture use (Zou *et al.*, 2013). Regarding bioplastic conversion, a mixed culture of different *Pseudomonas* has been studied to convert a mixture of benzene, toluene, ethylbenzene, *p*-xylene, and styrene, as a proxy for the aromatic fraction of plastic pyrolysis oil, into PHA (Nikodinovic *et al.*, 2008).

1.5.5 Simultaneous utilization of mixed substrates by a single strain

Adaptive laboratory evolution (ALE) is an adaptive mechanism conducted on a laboratory scale for prolonged periods, ranging from weeks to years, which allows the selection to obtain a specific phenotype. In the context of metabolic engineering, it purposes to study gene knockout responses, improve productivity via growth coupling, expand substrate ranges, and improve the tolerance of process conditions. Based on two latter advantages, adaptation is enormously essential in the biodegradation of complex monomers (Yang, 2018).

Nutrient sources and environmental stresses are two main aspects to regulate the adaptive mechanism of biodegradation. Microbial cells evolved mechanisms to cope with such stressful conditions for maintaining cellular homeostasis successfully. Therefore, preference and efficiency of nutrient source utilization are the key factors to establish proper ALE. For instance, recent data showed that *E. coli* could evolve different modes of lacZ expression, depending on the environment that contained either lactose, glucose, or both as carbon sources. Moreover, study ALE of *S. cerevisiae* showed that glucose-limited selection led to increased biomass yield and a decreased fermentative capacity (Dragosits and Mattanovich, 2013).

Temperature, salt, and organic solvent are considered important environmental stress factors that allow significant alterations of cells. For example, increased temperature leads to the mutation of genes accompanied by increased membrane fatty-acid saturation. Moreover, growth adaptation in high osmotic pressure could evolve *E. coli* to increase the expression of trehalose synthesis. Furthermore, ALE of organic solvent also significantly contributes to recent studies due to its promising future as next-generation biofuel production (Dragosits and Mattanovich, 2013).

Regarding ALE on PU monomers, Li (2019) reported that single strain *P. putida* KT2440 had been successfully evolved on adipic acid, 1,4-butanediol, and ethylene glycol. 1,4-butanediol and ethylene glycol utilization were obtained through adaptive laboratory evolution followed by reverse engineering, while adipic acid utilization was conducted via cloning of *dca* operon (*dcaAKIJP)* from *Acinetobacter baylyi* and further evolution.

1.6 Polyethylene terephthalate (PET) upcycling

Polyethylene terephthalate (PET) is a thermoplastic polyester with high stability in melting temperature (260°C) and has advantages in chemical, hydrolytic, and solvent resistance (Venkatachalam *et al.*, 2012). Due to those advantages, the polymer is extensively applied in industrial packaging and textile (Worm *et al.*, 2017). In 2015, PET global production contributed as much as 33 million metric tons, while 32 million tons of PET waste was generated from that number (Geyer *et al.*, 2017). The low recycling rate of PET affected the considerable amount of its waste in the environment, resulting in microplastic pollution. There is much evidence that microplastics have been accumulated at any level of the food chain. For examples, micro-plastics have been found in the gut content of sea birds and have been collected from seawater with plankton samples in North Sea (Barnes *et al.*, 2009).

As a strategy, chemical recycling is intensively combined with mechanical recycling to tackle much PET waste, such as in landfills. Chemical recycling can be defined as a process to convert polymers to monomers chemically. Hydrolysis is a popular method of chemical recycling that has been applied PET recycling. This polymer is reacted with water in an acid or alkaline or neutral environment leading to the depolymerization (Grigore, 2017).

1.6.1 Structure of PET: ethylene glycol and terephthalic acid

PET is commercially synthesized by reacting ethylene glycol (EG) either with (1) terephthalic acid (TA) or (2) dimethyl terephthalate (DMT). The reaction would yield bis(hydroxyethyl) terephthalate (BHET) as an intermediate product. After these initial reactions, the first polymerization step is transesterification between BHET and displacing EG. After the first polymerization, the resulted polymer can be subjected to further solid-state polymerization to obtain the required molecular weight (Webb *et al.*, 2012).

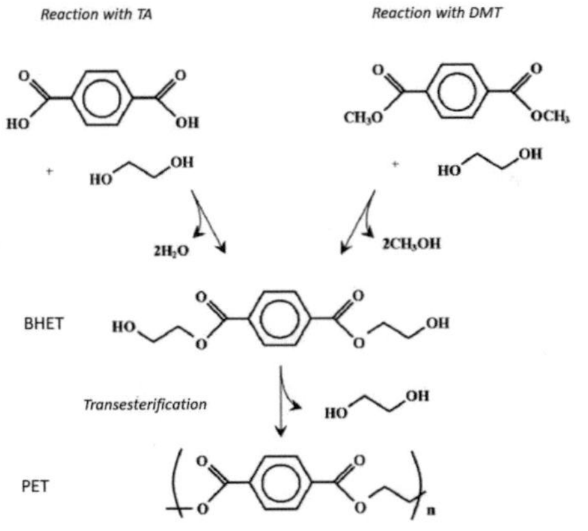

Figure 4 Chemical reaction of PET manufacturing either between EG and TA or EG and DMT (adapted from Webb et al., 2012)

1.6.2 PET enzymatic depolymerization

Surprisingly, fungi and bacterial polyester hydrolase exhibited hydrolytic activity of PET films and fiber. The microbial polyester hydrolase can depolymerize PET to ethylene glycol and terephthalic acid, which can be reused to synthesize virgin PET or utilized for PHA production. The enzymatic activity can optimally work on PET's transition temperature at 70°C while prohibited on PET's melting point at 230°C. On that point, the amorphous regions of PET become more flexible and accessible to enzyme attack. In this case, temperature as one of the abiotic factors can affect the effectiveness of enzyme penetration (Wei and Zimmermann, 2017a; Wei *et al.*, 2019) .

The other report also showed that a novel bacterium called *Ideonella sakaiensis* 201-F6 expressed PET hydrolase (PETase) activity on PET film degradation. PETase somehow is heat-labile but more active than the other enzymes, such as TfH, LCC, and FsC at 30°C. Even though the bio-recycling of PET at this temperature is unlikely to be applied, the natural degradation by this strain would be critical in most environments where PET products exist. During biodegradation of PET, the strain produces PETase and MHETase, in which PETase initially degrades it to mono(2-hydroxyethyl) terephthalic acid. Afterwards, the intermediate is degraded by MHETase to ethylene glycol and terephthalic acid (Yoshida *et al.*, 2016).

As an alternative to the hydrolases of *I. sakaiensis*, leaf-branch compost cutinase (LCC) has been intensively investigated due to its ability of PET biodegradation at >70 °C. The encoding gene of this polyester hydrolase was isolated from a plant compost metagenome (Sulaiman *et al.*, 2012). During the experiment, Tiso *et al.* received 215.6 mM terephthalic acid and 250.4 mM ethylene glycol from 15.7 g PET film after 120 h LCC-catalyzed hydrolysis in a stirred tank reactor (Tiso *et al.*, 2021). An optimized LCC regarding PET depolymerization was created by improving the enzymatic activity through targeted mutagenesis of amino acid residues. Furthermore, thermostability was enhanced through the addition of a disulfide bridge (Tournier *et al.*, 2020).

In addition to PET enzymatic depolymerization, the recombinant *Thermobifida fusca* cutinase (TfCut2) expressed in *Bacillus subtilis* could degrade PET packaging material as indicated by up to 50% weight loss after incubation at 70°C for 96 h. During ester cleavage, C=O and C-O bonds are expected to be broken. As a result of enzymatic activity, the abundance of ethylene glycol groups increases, and mid-chain methylene bonds decrease (Vague *et al.*, 2019). The recovery of monomers further could be followed by distillation to regain ethylene glycol and by filtration to obtain terephthalic acid due to its precipitation in the presence of a strong mineral acid (Carta *et al.*, 2003; Wei *et al.*, 2019).

1.6.3 Ethylene glycol metabolism

Ethylene glycol has a wide application in PET manufacture and as starting material in wide application, such as in the anti-freeze industry and chemical production of glyoxal, glycolic acid, and glyoxylic acid (Mattioda and Christidis, 2000). Ethylene glycol utilization by bacteria has also been intensively studied due to its promising circular bioeconomy from PET waste to value-added product. Therefore, Li (2019) investigated the pathway of ethylene glycol and evolved *Pseudomonas putida* KT2440 in ethylene glycol utilization. That study also demonstrated the reverse engineering to prove that mutation in the *gclR* gene indicates the significant growth of *P. putida* KT2440 on ethylene glycol as a sole carbon source, while the secondary mutations in PP_2046 and PP_2662 likely affect the rates of the initial oxidation reactions (Li *et al.*, 2019). In addition, the growth of the same strain got improved on ethylene glycol, when *gcl* operon was completely expressed (Franden *et al.*, 2018).

In *Pseudomonas putida* KT2440, the diol is converted into glyoxylate mainly by the dehydrogenases PP_0545, PedI, PedE and PedH (Mückschel *et al.*, 2012; Li *et al.*, 2019). This microbe also has the genetic ability (PP_4297-PP_4301) to metabolize glyoxylate into tartronate semialdehyde and CO_2 by glyoxylate carboligase (Gcl). The intermediate tartronate semialdehyde is directly converted to glycerate or via hydroxypyruvate followed by a conversion into 2-phosphoglycerate (Franden *et al.*, 2018). Li et al. 2019 revealed the role of the transcriptional regulator GclR in the Gcl pathway by analyzing *P. putida* KT2440 variants evolved to grow on ethylene glycol. This regulator plays a more extensive metabolic context of purine and allantoin metabolism and thus represses the PP_4297-PP_4301 gene cluster, which encodes for the Gcl pathway.

1.6.4 Terephthalic acid metabolism

Terephthalic acid degradation pathway was revealed in *Rhodococcus* sp., *Comamonas* sp., and *Pseudomonas umsongensis* GO16. Those can metabolize terephthalic acid into protocatechuate (PCA) through a pathway encoded by the *tph* operon. The gene cluster, ***tphCA2A3BA1,*** is responsible for terephthalic acid metabolism under the control of regulator ***tphR*** (see figure 5). During the investigation of *Comamonas* sp.

strain E6, terephthalic acid pathway has been observed that TphA2 (oxygenase large subunit of TPA 1,2-dioxygenase), TphA3 (oxygenase small subunit of TPA 1,2 - dioxygenase), and TphA1 (reductase component) are responsible for the degradation from terephthalic acid to DCD (1,2-dihydroxy-3,5-cyclohexadiene-1,4-dicarboxylate). TphB (1,4-DCD dehydrogenase) then metabolizes DCD to PCA (protocatechuate) before entering PCA 4,5-cleavage pathway (Sasoh *et al.*, 2006; Hara *et al.*, 2007; Kamimura *et al.*, 2010; Kasai *et al.*, 2010; Salvador *et al.*, 2019; Narancic *et al.*, 2021; Tiso *et al.*, 2021).

After that, PCA can be further degraded to intermediates of the TCA cycle via the protocatechuate 4,5-cleavage (PCA45) pathway (Wang *et al.*, 1995). Aromatic ring is further cleaved by PCA 4,5-dioxygenase. The product is then degraded to pyruvate and oxaloacetate via five further enzymatic steps (Kamimura *et al.*, 2010). Beside the degradation of phthalate isomers, the PCA45 pathway is also involved in the degradation of lignin-derived low-molecular- weight compounds (Sasoh *et al.*, 2006; Kamimura *et al.*, 2010). The previously mentioned bacteria, *Comamonas testosteroni* YZW-D, *Comamonas sp.* strain E6, *P. umsongensis* GO16, also express *pcaR* as a transcriptional regulator involved in terephthalic acid utilization (Wang *et al.*, 1995; Hosaka *et al.*, 2013; Narancic *et al.*, 2021). Furthermore, the difference among terephthalic acid degraders is that Rhodococci and *P. umsongensis* GO16 use the *p*-hydroxybenzoate transporter *pcaK* to transport terephthalic acid within the cell (Figure 5 B) (Hara *et al.*, 2007; Narancic *et al.*, 2021).

The latter mentioned, *P. umsongensis* GO16, was isolated from soil sample exposed to PET granules at PET bottle processing plant, is capable to utilize terephthalic acid as sole carbon source and known as PHA producer. The strain also showed the optimum and identical mechanism of terephthalic acid degradation by the activity of full operon of *pcaRtphA2A3B1A1pcaK* (Salvador *et al.*, 2019; Narancic *et al.*, 2021). In the purpose of *in situ* applications, most of Gram-negative microorganism, such as *Pseudomonas* sp. is used and released in environment as bioindicator and biodegrader. Therefore, the genetic scheme with Tn5- or Tn7- mini transposons targeting *att*Tn7 sequence would be significantly helpful to implant *tph* operon inside the genome of another *Pseudomonas* sp. for certain upcycling purpose (De Las Heras *et al.*, 2008). This single *att*Tn*7* site was specifically located close to *glmS*, which encodes the essential enzyme glucosamine-6-phosphate synthetase (Choi *et al.*, 2005; Y. J. Choi *et al.*, 2006).

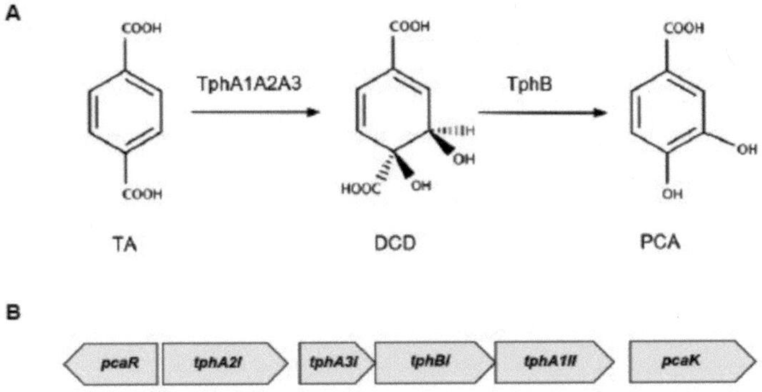

Figure 5 Terephthalic acid metabolism. A. Terephthalic acid degradation to protocatechuate in bacteria. **B.** Genetic organization of the *tph* operon in *Pseudomonas umsongensis* GO16 (adapted from Salvador *et al.*, 2019).

1.7 High-value products from plastic monomer upcycling

Rhamnolipids, phenazine, and polyhydroxyalkanoates are three of the most promising products in upcycling of plastic waste, which would significantly contribute on circular bioeconomy. Rhamnolipids are leading biosurfactant, which have a lot of applications in agricultural, cleansing, and pharmaceutical, and mostly found as ingredients in biopesticide, cosmetic, detergent, and medicine. Based on glucose two pathways are required for rhamnolipid synthesis. For production of mono-rhamnolipid, activated β-hydroxy-fatty acids are formed via fatty acid de novo synthesis, which are fused by the enzyme RhlA. Activated rhamnose is synthesized and subsequently coupled to the β-hydroxy-fatty acid dimer called hydroxyalkanoyloxy alkanoate (HAA) by the rhamnosyltransferase I (RhlB) (Tiso *et al.*, 2017, 2021; Utomo *et al.*, 2020).

Pseudomonas aeruginosa is considered as a native producer of phenazine having nine phenazine synthesis genes *phz A1-G1, phz M*, and *phz S*. Encoded enzymes Phz A1-G1 are responsible to convert chorismate from the shikimate pathway to phenazine 1-carboxylic acid (PCA). Afterwards, PCA is converted to 5-methylphenazine-1-carboxylic acid (5-MCA) by Phz M, and then converted by Phz S to the end-product, pyocyanin. The heterologous expression of those genes was successful to produce phenazine in *P. putida* KT2440 with glucose as a sole carbon source in bio-electrochemical system (Schmitz *et al.*, 2015; Askitosari *et al.*, 2019).

Pseudomonads are the promising groups to be engineered as mcl-PHA producer. As a residue monomer of polyethylene terephthalate and polyurethane, ethylene glycol can be utilized by engineered strains (e.g., *Pseudomonas putida* MFL185) as a carbon source. Furthermore, this strain can produce 32% mcl-PHA of its dry cell weight with a yield of 0.06±0.00 g of mcl-PHA produced per g of ethylene glycol (Franden *et al.*, 2018). PHA production was also observed on evolved *Pseudomonas putida* KT2440 with 1,4-butanediol as a sole carbon source; 0.12 g/L of the product was obtained (Li *et al.*, 2020). Interestingly, the PHA was also produced by other *Pseudomonas* species, namely *P. umsongensis* GO16, 19, and 23 from upcycled PET (Kenny *et al.*, 2008).

As an alternative of above-mentioned products, methyl ketone could be a desired product from plastic monomer upcycling. The overexpression of corresponding genes in fatty acid β-oxidation led the production of methyl ketone using glucose as a sole carbon source. The deletion of genes, namely *tesB, fad A-A2,* and *fadE-E2* led to the production of high titer methyl ketone. Once plastic monomers can be used as substrates, the plastic-to-biofuel process would be feasible (Nies *et al.*, 2020).

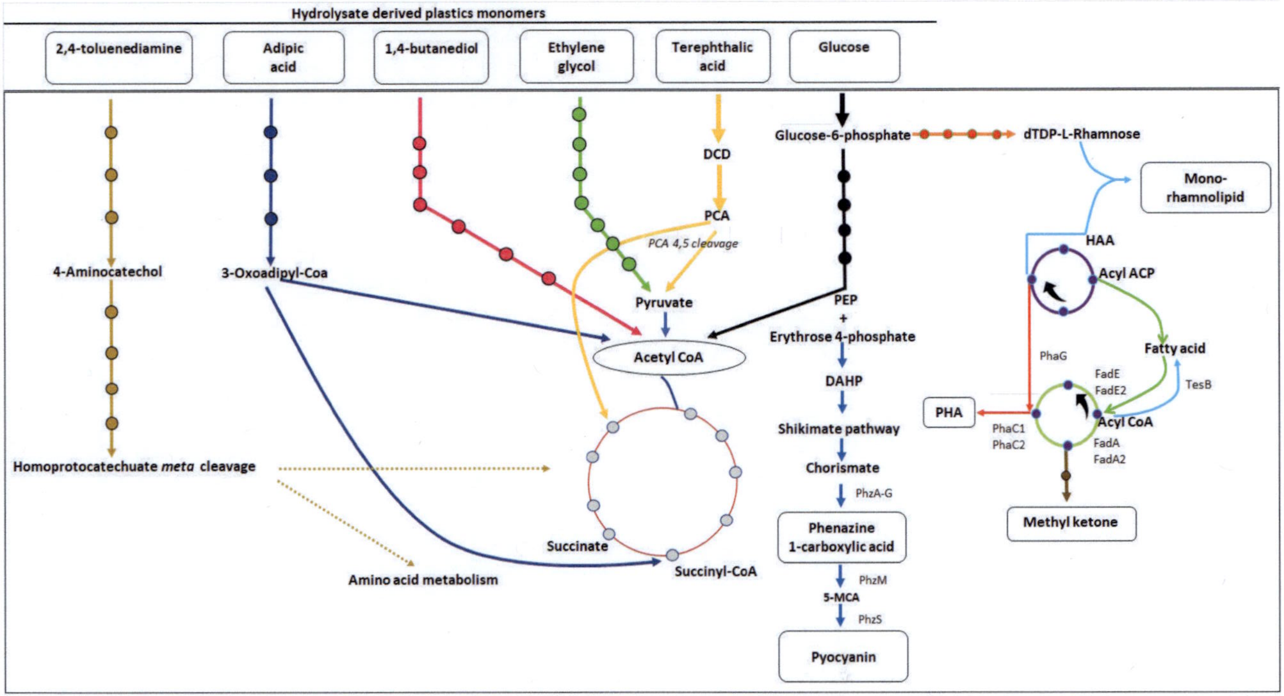

Figure 6. Metabolic pathway of plastic monomers and glucose as a reference substrate for production of valuable products, namely mono-rhamnolipid, phenazine, PHA, and methyl ketone.

1.8 Scope of thesis

The use of PU and PET monomers as carbon source for the production of valuable products is aimed at this study. This approach should highlight the possibilities of plastic waste upcycling and hence contribute to the current discussion of implementing a sustainable circular plastic economy.

In Chapter 3.1, PU monomers were used as sole carbon source by a defined mixed culture. Each monomer was utilized by a specialist microbe. Growth inhibition by 2,4-toluenediamine was overcome by the prior extraction of this valuable compound. Finally, the mixed culture was used in proof-of-principle experiments for growth and bio-detergent production from mixed PU monomers.

Chapter 3.2 is designated to simultaneous PET monomers utilization by *P. capeferrum* TDA1 for rhamnolipid synthesis. Several strategies, including ALE, genomic integration, further ALE in mini bioreactors, and pH optimization, were applied for the improvement of simultaneous utilization of terephthalic acid and ethylene glycol.

To follow up, chapter 3.3 is about the use of *P. capeferrum* TDA1 for the utilization of PU monomers and the production of rhamnolipids. Since the strain is recognized as a 2,4-TDA utilizer, the reactive extraction was omitted in this study. However, the toxicity of increased 2,4-TDA concentrations in mineral salt medium (MSM) was also investigated to identify optimal conditions for growth and rhamnolipid production.

For diversification of products, chapter 3.4 focuses on phenazine production as an antifungal compound by a dedicated strain of *P. capeferrum* TDA1 from PU or PET monomers as sole carbon sources. The toxicity test of phenazine on *Saccharomyces cerevisiae* as a model was also conducted

The exploitation of the bow-tie structure from different substrates to rhamnolipids and phenazines is reported in Chapter 3.5. The selected monomers for the bow-tie structure represents different polymers mimicking household waste or the mixture of waste in the environment.

Chapter 3.6 reports on *P. capeferrum* TDA1 as a native bioplastic (PHA) producer. Plastic monomers and the other alternative substrates were investigated as carbon source.

To get closer to industrial application, the characteristic of *P. capeferrum* TDA1 as S1 strain is one of advantages. In chapter 3.7, the strain was tested to see another advantage as a tolerant strain at elevated temperatures.

Chapter 2

Materials and Methods

Contributions:

This chapter was written by Romualdus Nugraha Catur Utomo and reviewed by Lars M. Blank. Some parts of this chapter were also published and could be found in:" Defined Microbial Mixed Culture for Utilization of Polyurethane Monomers." *ACS Sustainable Chemistry & Engineering* 2020, 8, 47, 17466-17474 and in:" New species *Pseudomonas capeferrum* TDA1 as a plastic monomer utilizer and a PHA native producer." *IOP Conf. Series: Earth and Environmental Science* 2021, 894, 012024.

2.1 Chemicals and materials

All monomers, namely adipic acid, 1,4-butanediol, ethylene glycol, 2,4-toluenediamine, terephthalic acid, were purchased from Sigma-Aldrich Chemie GmbH. Sulfuric acid (98%) for the HPLC buffer was purchased from Carl Roth GmbH & Co. KG (Karlsruhe, Germany). Dichloromethane for GC-sample preparation was purchased from Carl Roth GmbH & Co. KG (Karlsruhe, Germany. Paraffin oil and di-(2-ethylhexyl) phosphoric acid (D2EHPA) were used as a solvent and reactive extractant (co-solvent) during extraction and were purchased from Carl Roth GmbH & Co. KG (Karlsruhe, Germany) and Sigma-Aldrich Chemie GmbH, respectively.

2.2 Microorganisms and culture conditions.

All bacteria used in this study are listed in the **Table 2** and cultivated in LB medium consisting of 10 g/L peptone, 5 g/L yeast extract, and 5 g/L sodium chloride or in mineral salt medium (MSM) (Hartmans *et al.*, 1989) using the different plastic monomers or monomer mixture as carbon source(s) as indicated.

Table 2 Bacterial strains used in this study with listed genotype and references

iAMB strain number	Strain	Resistance	Genotype	Ref.
3045	*Acinetobacter baylyi*	-	DSM No. 24193	(Parke *et al.*, 2001)
Pseudomonas putida KT2440				
4631	A12.1p	-	ALE on adipic acid, cured	(Li, 2020; Ackermann *et al.*, 2021)
3566	B 10.1	-	Grow on 1,4-butanediol	(Li *et al.*, 2020)
4647	Δ*gclR* ΔPP_2046 ΔPP_2662::14d	-	Grow on ethylene gylcol	(Li *et al.*, 2019)
6208	A12.1p pPS05	Tet10	Grow on adipic acid and harboring *rhlAB* for mono-rhamnolipid production	(Utomo *et al.*, 2020)
6210	B 10.1 pPS05	Tet10	Grow on 1,4-butanediol and harboring *rhlAB* for	(Utomo *et al.*, 2020)

			mono-rhamnolipid production	
6211	Δ*gclR* ΔPP_2046 ΔPP_2662::14d pPS05	Tet10	Grow on ethylene glycol and harboring *rhlAB* for mono-rhamnolipid production	(Utomo *et al.*, 2020)
Pseudomonas capeferrum TDA1				
4916	Wild type	-	Strain utilizing 2,4-TDA	(Cárdenas Espinosa *et al.*, 2020)
6217	ALE TDA1	-	ALE on PU monomers	
7225	ALE TDA1p_ *tph*	Km50	ALE TDA1 with pBT'T harboring *tph* operon	
6218	ALE TDA1ge_*tph*	-	ALE TDA1 with integrated tph operon	
6219	ALE TDA1ge_*tph* F1.1	-	ALE TDA1ge_*tph* evolved in mini bioreactor	
6221	ALE TDA1ge_*tph* F1.1T	-	ALE TDA1ge_*tph* F1.1 with turbidostatic feeding system	
7159	ALE TDA1ge_*tph* F1.1 + pPS05	Tet10	ALE TDA1ge_*tph* F1.1 with pPS05 harboring *rhlAB* for mono-rhamnolipid production	
7160	ALE TDA1ge_*rhlAB*	-	ALE TDA1 with integrated *rhlAB* for mono-rhamnolipid production	
7161	ALE TDA1ge_*rhlAB*_evAA	-	ALE TDA1 ge_*rhlAB* with pBNT_mcs dcaAKIJPcpi, cured	
7162	ALE TDA1evAA + pJNN_*phzAG_phzMS*	Gm25/ Amp100	ALE TDA1 with pBNT_mcs dcaAKIJPcpi, cured, and transformed with pJNN harboring *phzA1-G* and *phzMS*	

7163	ALE TDA1p_*tph* + pJNN_*phzAG_phzMS*	Kan50 + Gm25/ Amp100	ALE TDA1p_ *tph* with pJNN harboring *phzA1-G* and *phzMS*	
7164	ALE TDA1ge_*rhlAB* + p_ *tph*	Kan50	ALE TDA1 ge_*rhlAB* with pBT'T *tph* operon	
Escherichia coli				
	DH5α λpir pPS05	Tet10	pPS05 harboring *rhlAB*	(Tiso *et al.*, 2016)
	pir2 *tph* operon	Gm25	Tn7 donor of tph operon	
2037	HB101 pRK2013	Km50	Helper strain	
3221	pTnS-1	Amp100	Expressing transposase	

Pseudomonas and *A. baylyi* were incubated at 30°C, while *Escherichia coli* was grown at 37°C. For pre-culture, the strains were inoculated into 2 ml MSM in 24-well System Duetz plates (Enzyscreen, Heemstede, The Netherlands) or 10 ml MSM in 100 ml shake flask, cultivated in a Multitron shaker (INFORS, Bottmingen, Switzerland) with a 300 rpm or 200 rpm shaking speed with an amplitude of 50 mm. For main culture, the strains were grown in 250 ml shake flasks with a 25 ml MSM filling volume using Multitron shaker (INFORS, Bottmingen, Switzerland). The liquid cultivations were incubated at 30°C and 200 rpm shaking speed using the different plastic monomers as a carbon source or unless specifically stated.

For online growth detection, the strains were cultured in a Growth Profiler 960 (Enzyscreen, Hemmstede, The Netherlands) using 24 well microtiter plates (2 ml MSM). Cell densities are expressed as green pixel value (G value), derived from image analysis of microtiter plates with transparent bottoms. The liquid cultivations were incubated at 30°C and 225 rpm shaking speed using the different plastic monomers as a carbon source.

2.3 Adaptive laboratory evolution

Adaptive laboratory evolution (ALE) was performed as follows: a pre-culture of *P. capeferrum* TDA1, cultivated in MSM with 20 mM adipic acid, was used to inoculate 250 mL clear glass Boston bottles with Mininert valves (Thermo Fisher Scientific, Waltham, MA, USA) containing various compositions for the adaptation on ethylene glycol. Serial transfers were reinoculated several times after the cultures reached an OD_{600} of at least 0.5, with a starting

OD_{600} of 0.1. After growth was detected (usually overnight), single colonies were isolated from ALE cultures by streaking samples on LB agar plates. The best mutant was chosen out of 10 tested colonies according to the highest growth rate and biomass in MSM with 30 mM ethylene glycol as a sole carbon source determined using the Growth Profiler 960 (Enzyscreen, Heemstede, The Netherlands).

2.4 Molecular engineering

2.4.1 Transformation

For plasmid-based growth on terephthalic acid, the *tph* operon from *Pseudomonas putida* GO16 has been cloned and constructed into pBT'T plasmid as a vector. After that, the plasmid was transformed into TDA1 via electroporation (Choi *et al.*, 2006). For obtaining the positive clones, the strain was spread on LB agar + kanamycin 50 mg/L and screened in MSM with 20 mM terephthalic acid as a sole carbon source using a 24-deep well plate in a Multitron shaker.

2.4.2 Tn7 integration system

The constructed plasmid (pBG_pcaRtphA2A3B1A1pcaK) harboring the *tph* operon in the Tn7-mini transposon was transformed for the integration-based approach to *Escherichia coli* competent cell via electroporation (Choi *et al.*, 2006). The integration of the *tph* operon into *Pseudomonas* strains was conducted via parental mating between the Tn7 donor (*Escherichia coli* pir2 *tph* operon), recipient (*Pseudomonas* strains), helper (*Escherichia coli* HB101 pRK2013, and pTnS-1 (transposase)-harboring *Escherichia coli* on LB agar at 30°C overnight. As *Pseudomonas* selection, the mating was streaked on cetrimide agar with gentamicin 30 mg/L as a selective antibiotic for Tn7 construction. For the selection, at least 10 single colonies were inoculated on MSM containing terephthalic acid 20 mM using a 24-deep well plate in a Multitron shaker.

2.4.3 Genome sequencing analysis

Genomic DNA of *P. capeferrum* TDA1 and the evolved variants for resequencing were isolated through a High Pure PCR Template Preparation Kit (ROCHE life science, Basel, Switzerland). Sequencing was done by Eurofins Genomics (Ebersberg, Germany) using genome sequencer Illumina HiSeq with 2 x 150 bp paired - end read mode. Genome variant analysis includes mapping against one reference, detecting single nucleotide polymorphisms

(SNPs) and insertion deletions (InDels) (using dbSNP), annotating detected SNPs and InDels, and allocating protein levels. Analyzed SNPs and InDels were listed and visualized with the Integrative Genomics Viewer (IGV) (McKenna *et al.*, 2010; DePristo *et al.*, 2011; Thorvaldsdottir *et al.*, 2013).

2.4.4 Reverse transcription quantitative PCR

The RT-qPCR is a sensitive method allowing the direct verification and quantification of gene transcript (Udvardi *et al.*, 2008). For RT-qPCR experiment, the CAPITAL qRT-PCR Green Mix (biotechrabbit, Germany) was used. The thermocycling condition and master mix preparation for DNA and RNA as follows:

Table 3 Master mix of CAPITAL qPCR Green Mix for DNA

Component	Volume
CAPITAL qPCR Green Mix 4x	2.5 µl
10 µM forward primer	0.5 µl
10 µM reverse primer	0.5 µl
Template	1.0 µl
H_2O	5.5 µl
Total	10 µl

Table 4 Master mix of CAPITAL qPCR Green Mix for RNA

Component	Volume
CAPITAL qPCR Green Mix 4x	2.5 µl
RTase with RNAse inhibitor 20x	0.5 µl
10 µM forward primer	0.5 µl
10 µM reverse primer	0.5 µl
Template	1.0 µl
H_2O	5.0 µl
Total	10 µl

Table 5 Thermocycling protocol of RT-qPCR with DNA

Step	Temperature	Duration	Cycle
Initial denaturation	95°C	3 min	1
Denaturation	95°C	10 sec	45
Annealing	64°C	30 sec	45

Table 6 Thermocycling protocol of RT-qPCR with RNA

Step	Temperature	Duration	Cycle
Reverse transcription	50°C	10 min	1
Initial denaturation	95°C	3 min	1
Denaturation	95°C	10 sec	45
Annealing	64°C	30 sec	45

Analysis of RT-qPCR data was performed using CFX Manager software (Bio-Rad, USA). The generated data were normalized with three statistical methods:

1. Livak method(Livak and Schmittgen, 2001)

The first step was to calculate the difference between cycle threshold (C_t) values between the target gene (TG) and housekeeping gene (HKG).

$$\mathbf{\Delta Ct\ (TG) = Ct\ (TG) - Ct\ (HKG)}$$

For further calculation, establish the average of ΔCt of control sample (CS) in this case *gntR*-non expressing samples.

$$\mathbf{\Delta Ct\ (CS) = Ct\ (CS) - Ct\ (HKG)}$$

Afterwards, ΔΔCt was obtained by calculating the difference of ΔCt (TG) for *gntR*-expressing samples and ΔCt (CS) for *gntR*-non expressing samples.

$$\mathbf{\Delta\Delta Ct = \Delta Ct\ (TG) - \Delta Ct\ (CS)}$$

The expression fold change was obtained by two to the power of the negative of ΔΔCt.

$$\textbf{Expression fold change} = \mathbf{2^{-(\Delta\Delta Ct)}}$$

2. Pfaffl method(Pfaffl, 2001)

This method takes the primer efficiencies into consideration. Primer efficiencies (E) for each experiment were calculated using the LineRegPCR software (Ramakers *et al.*, 2003; Ruijter *et al.*, 2009). Initially, an average Ct of TG and CS was determined then subtracted by the raw Ct from corresponding sample.

$$\mathbf{\Delta Ct\ (TG) = avg\ Ct\ (TG) - Ct\ (TG_n)}$$

$$\Delta Ct\ (CS) = avg\ Ct\ (CS) - Ct\ (CS_n)$$

$$\Delta Ct\ (HKG) = avg\ Ct\ (HKG) - Ct\ (HKG_n)$$

Afterwards, the gene expression ratio was calculated

$$\text{Gene expression ratio} = \frac{(E_{TG})^{\Delta Ct\ (TG)}}{(E_{HKG})^{\Delta Ct\ (HKG)}}$$

$$= \frac{(E_{CS})^{\Delta Ct\ (CS)}}{(E_{HKG})^{\Delta Ct\ (HKG)}}$$

3. Multiple HKG analysis (Vandesompele *et al.*, 2002; Hellemans *et al.*, 2007)

This analysis is related to the Pfaffl method because it considers not only the primer efficiencies, but also the multiple HKG. Primer efficiencies (E), the average Ct values (avg Ct), and ΔCt values were calculated previously in the Pfaffl method. For further step, the relative quantity (RQ) values for each gene were determined.

$$RQ = E^{\Delta Ct}$$

Thereafter, the geometric mean of the RQ values of the HG was formed.

$$\text{Geometric mean } (RQ_{HKG}) = \sqrt[n]{HKG1\ .\ HKG2\ .\ HKGn}$$

Finally, the relative gene expression was calculated by dividing the RQ of TG by the geometrical mean of the RQ of the HKG.

$$\text{Relative gene expression} = \frac{RQ_{TG}}{\text{Geometric mean } (RQ_{HKG})}$$

2.5 Adaptive laboratory evolution with mini bioreactor

Further adaptive laboratory evolution was also conducted using an in-house mini bioreactor connected to a PC with real-time growth rate measurement. The working volume of the fermenter was 12 ml with initial OD 0.1 and was followed by a sequential feeding system. Separated monomer feeding (first 20 mM terephthalic acid and next 30 mM ethylene glycol) was applied in this study. After reaching a certain growth rate on the targeted monomer, single colonies were isolated from ALE cultures by streaking samples on LB agar plates. One best colony (ALE TDA1ge_*tph* F1.1) was selected according to the growth behavior in MSM with 20 mM terephthalic acid and 30 mM ethylene glycol determined using the Growth Profiler 960 (Enzyscreen, Heemstede, The Netherlands).

2.6 Analytical methods

2.6.1 Monomer analysis

Three monomers, adipic acid, 1,4-butanediol, and ethylene glycol, were analyzed by High-Performance Liquid Chromatography (HLPC) using a Beckman System Gold 126 Solvent Module equipped with autosampler, column oven (Beckman Coulter GmbH, Krefeld, Germany), and a Smartline 2300 refractive index detector (Knauer, Berlin, Germany). Analytes were separated using a 300 x 7.8 mm, Metab -AAC, BF-Series column (Isera, Düren, Germany) and a 40 x 8 mm organic acid resin precolumn (CS Chromatographie, Langerwehe, Germany). The eluent was 5 mM H_2SO_4 at a flow rate of 0.7 ml/min. Glucose used in bow-tie-exploiting experiment was analyzed with the same method.

Terephthalic acid was analyzed by High-Performance Liquid Chromatography (HLPC) using a Beckman System Gold 126 Solvent Module equipped with a UV detector. Analytes were eluted using 150 x 4 mm isopher 100-05 C18 column (Knauer, Berlin, Germany). The eluents were acetonitrile and 1% trifluoroacetic acid with a gradient system at a flow rate of 0.8 ml/min.

2,4-toluenediamine was quantified by Gas Chromatography (GC) equipped with Flame Ionization Detector (FID) (Thermo Scientific-Trace GC Ultra). A ZB-waxplus column was used with helium as a carrier gas. For quantification, 2,4-toluenediamine was extracted using dichloromethane with a sample to dichloromethane ratio of 2:1 (0.4 ml sample + 0.2 ml dichloromethane). Afterwards, the solution was vortexed for 30 mins at 400 rpm, centrifuged for 5 mins at 13,000 rpm, and the lower phase was analyzed by GC-FID. The split ratio was set to 20 according to the manual of Dikma Technologies Inc. Decanoate used in bow-tie-exploting experiment was also quantified with the same method using GC-FID, while a sample was prepared by mixing with methanol with ratio of 1:1 (0.2 ml sample + 0.2 ml methanol). Afterwards, the solution was thoroughly vortexed, centrifuged for 2 mins at 13,000 rpm, the supernatant filtered, and injected to GC-FID (Dikma Technologies Inc, 2018).

2.6.2 Extraction of 2,4-toluenediamine from the microbial fermentation broth

The model PU hydrolysate, consisting of 121 mM adipic acid, 60 mM 1,4-butanediol, 60 mM ethylene glycol, and 8.5 mM 2,4-toluenediamine, was dissolved in mixed dipotassium phosphate and monosodium phosphate buffer (36 mM) that was pH adjusted (see results and discussion for details). As a reactive extractant, 5 wt% D2EHPA in paraffin oil was used in a medium to extractant ratio of 1:1. Afterward, the emulsion was rotated for 90 mins at 30°C at 90 rpm in a rotating contactor. Then, samples were centrifuged for 5 mins at 4,000 rpm to separate the organic and aqueous phases. The aqueous phase was adjusted to pH 7 (otherwise would be specifically mentioned), and filtered for HPLC-RI and GC-FID analyses. For complementing the buffer to a microbial growth medium, trace elements, ammonium sulfate (15.1 mM), and buffer (36 mM) were added into the aqueous phase (Utomo *et al.*, 2020), otherwise would be specifically mentioned.

2.6.3 Rhamnolipid production and analysis

For mono-rhamnolipid production, plasmid pPS05 harboring *rhlA* and *rhlB* was transformed into the *Pseudomonas* strains via electroporation (Choi *et al.*, 2006; Tiso *et al.*, 2016). The recombinant strains were further cultured in MSM with either a single monomer or mixed monomer mixture as a carbon source, and tetracycline was added as a selective antibiotic.

Mono-rhamnolipids were quantified with reversed phase HPLC (Behrens *et al.*, 2016). In brief, the rhamnolipid congeners were separated on a reversed-phase chromatography (C18 column, Macherey-Nagel) at 40°C and detected by a corona-charged aerosol detector (CAD) (Thermo Fisher). Acetonitrile and 0.2% formic acid were used as a mobile phase at a flow rate of 1 ml/min with the gradient system (Tiso *et al.*, 2017).

2.6.4 Phenazine production and analysis

Plasmid pJNN harboring *phzA1-G1, phzM,* and *phzS* was transformed into the *Pseudomonas* strains via electroporation for phenazine production. The recombinant strains were further cultured in MSM with either a single monomer or mixed monomer mixture as a carbon source, and gentamycin was added as a selective antibiotic.

Phenazine was measured by high-performance liquid chromatography (HPLC) with a C18 column (Waters Corporation) equipped with a photodiode array UV/VIS detector (Thermo Fisher), which was detected and quantified at 366 nm for phenazine-1-carboxylic acid (PCA) and at 280 nm for pyocyanin (PYO). Separation was achieved with a gradient of 0.1% trifluoroacetic acid (TFA) in acetonitrile (solution A) and 0.1% TFA in water (solution B) as mobile phase (with solution A at 15% for 2 min, 100% for 15 min, and 5% for 3 min) at a flow rate of 0.8 mL/min and a column temperature of 20°C (Askitosari *et al.*, 2019). Analytical grade PCA (Apollo) and PYO (Cayman Chemical) were used as standard solutions.

2.6.5 Polyhydroxyalkanoates staining and microscopy observation

For qualitative analysis of PHA by staining, 0.2% of a stock solution of 0.25 mg Nile red (Sigma-Aldrich Chemie GmbH) per ml dimethylsulfoxide (DMSO) was added into the sterilized medium to give a final concentration of 0.5 µg/ml. The agar plates were exposed to ultraviolet light (312 nm) after appropriate cultivation periods to detect the accumulation of PHAs (Spiekermann *et al.*, 1999). For qualitative analysis by microscopy observation, the culture from appropriate cultivation periods was observed under Leica DM1000 Led microscope using dark field (DF) contrast with 1,000 folds of magnification.

Chapter 3

Results

3.1 Defined microbial mixed culture for utilization of polyurethane monomers

Romualdus Nugraha Catur Utomo[1], Wing-Jin Li[1], Till Tiso[1], Christian Eberlein[2], Moritz Doeker[3], Hermann J. Heipieper[2], Andreas Jupke[3], Nick Wierckx[1,4], Lars M. Blank[1]

[1]Institute of Applied Microbiology - iAMB, Aachen Biology and Biotechnology – ABBt, RWTH Aachen University, Aachen, Germany

[2]Department of Environmental Biotechnology, Helmholtz Centre for Environmental Research – UFZ, Leipzig, Germany

[3]AVT, Fluid Process Engineering, RWTH Aachen, Aachen, Germany

[4]Forschungszentrum Jülich, Institute of Bio- and Geosciences IBG-1: Biotechnology, Jülich, Germany.

Published partially as:" Defined Microbial Mixed Culture for Utilization of Polyurethane Monomers." *ACS Sustainable Chemistry & Engineering* 2020, 8, 47, 17466-17474. DOI: https://doi.org/10.1021/acssuschemeng.0c06019.

Contributions:

1. This chapter was written by Romualdus Nugraha Catur Utomo and revised by Lars M. Blank and Nick Wierckx. Romualdus Nugraha Catur Utomo performed cultivation of dedicated strains, toxicity test, extraction process, and rhamnolipids production.
2. Wing Jin Li provided strain *P. putida* KT2440 A12.1p, B10.1, and Δ*gclR* ΔPP_2046.
3. Christian Eberlein and Hermann J. Heipieper provided strain *P. capeferrum* TDA1.
4. Till Tiso provided plasmid pPS05 for rhamnolipid production.
5. Moritz Doeker and Andreas Jupke gave an insight of extraction process.
6. This chapter was reviewed by Wing Jin Li, Till Tiso, Christian Eberlein, Moritz Doeker, Hermann J. Heipieper, and Andreas Jupke.

3.1.1 PU monomer utilization by dedicated microbes

We aimed to construct a defined microbial mixed culture for the conversion of a representative mock PU hydrolysate to valuable products. Therefore, we initially identified microbes that could utilize the single PU monomers adipic acid, 1,4-butanediol, ethylene glycol, and 2,4-toluenediamine as a sole carbon and energy source for growth.

Table 7 PU Monomer utilization by selected microbes (Utomo *et al.*, 2020)

Microbes	growth on single monomer*			
	adipic acid	**1,4 – butanediol**	**ethylene glycol**	**2,4-toluenediamine**
***Acinetobacter baylyi* ADP1**(Parke *et al.*, 2001)	++++	+	-	-
***P. putida* KT2440 A12.1p**(Li, 2020)	++++	-	-	-
***P. putida* KT2440 B10.1**(Li *et al.*, 2020)	-	++++	-	-
***P.putida* KT2440 Δ*gclR* ΔPP_2046**	-	-	+++	-
***P. capeferrum* TDA1**(Cárdenas Espinosa	+++	++	-	+

*: Growth was assessed with 60 mM carbon equivalent of specific monomers as sole carbon sources (adipic acid: 10 mM; 1,4-butanediol: 15 mM; ethylene glycol: 30 mM; 2,4-toluenediamine 8.5 mM) in minimum mineral salt medium (MSM) at 30°C by Growth Profiler in 24-square-well plates. Growth: ++++ to an OD > 2 in 24 hours, +++ to an OD > 1.5 in 24 hours, ++ to an OD > 1 in 60 hours, + to an OD > 1 in 168 hours, - no growth.

As described previously, *Acinetobacter baylyi* ADP1 grew well on adipic acid (Parke *et al.*, 2001). *P. putida* KT2440 A12.1p is an engineered strain harboring the *dca* operon of *A. baylyi* (*dcaAKIJP*), which was evolved for faster growth on adipic acid (Li, 2020). This strain grew as well as *A. baylyi* on adipic acid; hence, *P. putida* KT2440 A12.1p was used in further experiments to facilitate subsequent cultivation and strain engineering. *P. putida* KT2440 B10.1 (Li *et al.*, 2020) and *P. putida* KT2440 Δ*gclR* ΔPP_2046 ΔPP_2662::14d (Li *et al.*, 2019) were also selected for substrate specialization, which grew well on 1,4-butanediol, and ethylene glycol, respectively (table 7). *P. capeferrum* TDA1 was used for 2,4-toluenediamine degradation since it was reported to degrade 2 mM of 2,4-toluenediamine within approximately one week (Cárdenas Espinosa *et al.*, 2020). In a previous study, the degradation of 0.55 mM 2,4-toluenediamine required up to 4 days and ethanol as a co-substrate in an undefined mixed culture (Freedman DL, Shanley RS, 1996) highlighting the individual catalytic performance of *P. capeferrum* TDA1. With strains in hand that could utilize the four PU monomers, we initially designed a defined mixed culture consisting of the four *Pseudomonas* strains (table 7), which is subsequently characterized. Furthermore, a defined mixed culture of *Pseudomonas* strain would boost the engineering of product formation due to well-proved plasmid compatibility. In general, *Pseudomonas* also has the exceptional ability due to its organic solvent tolerance and its robustness as host in adaptive laboratory evolution for the conversion of aromatic compound to biofuels and bioproducts (Hosseini *et al.*, 2017; Mohamed *et al.*, 2020).

3.1.2 Biocompatible extraction of 2,4-toluenediamine

A mixture of 10 mM adipic acid, 15 mM 1,4-butanediol, and 30 mM ethylene glycol (60 mM carbon equivalent of each) yielded better growth of the mixed culture than the same mixture with the addition of 8.5 mM 2,4-toluenediamine. Without 2,4-toluenediamine, the mixed culture could completely utilize all three monomers within 24 h, while it required 48 hours and reached a lower OD_{600} in the presence of 2,4-toluenediamine, and even 7.4 mM of 2,4-toluenediamine still remained, confirming its inhibitory activity.

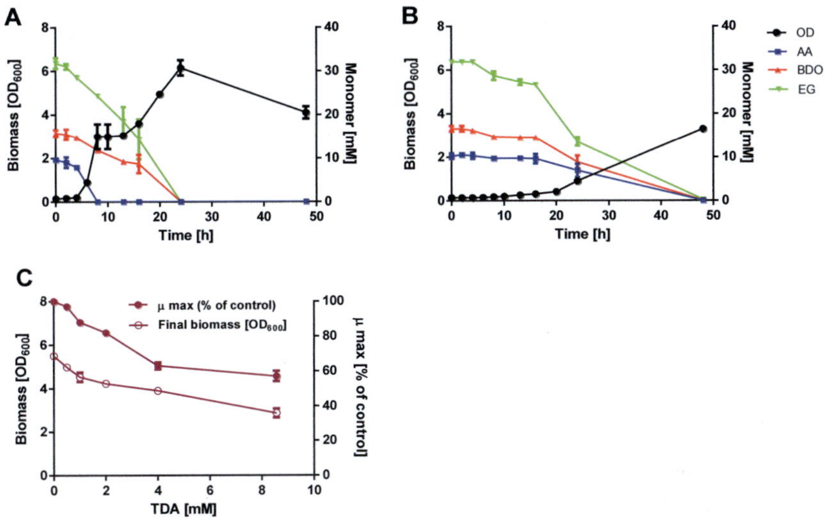

Figure 7 Growth and substrate concentration of a defined mixed culture of four Pseudomonas strains in MSM with PU monomer mixture consisting of **A:** 10 mM AA, 15 mM 1,4-BDO, and 30 mM EG and **B:** the same mixture which additionally includes 8.57 mM 2,4-TDA. **C:** Toxicity test of 2,4-TDA on a defined mixed culture of four Pseudomonas strains in MSM with PU monomer mixture consisting of 10 mM AA, 15 mM 1,4-BDO, and 30 mM EG and supplemented with different concentration of 2,4-TDA. Filled circles: effect of TDA on growth rate of a defined mixed culture, and empty circles: effect of 2,4-TDA on biomass of a defined mixed culture. The monomer mixture without 2,4-TDA was used as a control. Errors depict the deviation of the mean (n = 2) (Utomo *et al.*, 2020).

The toxicity test of 2,4-toluenediamine and its effect of *P. putida* KT2440 have been observed by adding 2,4-toluenediamine to exponentially growing cells with succinate as an initial carbon and energy source. In the presence of 2 mM 2,4-toluenediamine, the growth rate was reduced to 55% (Cárdenas Espinosa *et al.*, 2020). Toxicity test was also conducted in this study to investigate 2,4-toluenediamine effect on a defined mixed culture growing in MSM with PU monomer mixture as sole carbon sources.

Toxicity test has shown that the higher concentration of 2,4-toluenediamine the lower growth rate and biomass were obtained (see figure 7). Given the high toxicity of 2,4-toluenediamine, its extraction from the mock PU hydrolysate would likely enhance the mixed culture's growth. Furthermore, the extracted and recycled 2,4-toluenediamine could be used for TDI re-synthesis and subsequently for PU synthesis or further degradation by *P. capeferrum* TDA1 strain (European Union Risk Assessment Report, 2008; Cárdenas Espinosa *et al.*, 2020). In addition, the reactive extractant could be recycled and reused for the next extraction via back extraction (see figure 10) (Song *et al.*, 2015). Bednarz et al. reported a reactive extraction method for diamines from a microbial growth medium that was temperature-insensitive, while pH-sensitive. They focused on aliphatic hexane- and butane diamines, which, contrary to the aromatic diamines, have a pKa value in the basic range. Here, we adapted the method for 2,4-toluenediamine extraction. As reactive extractant, di-(2-ethylhexyl)phosphoric acid (D2EHPA) was used in paraffin oil as a carrier and biocompatible solvent (Bednarz *et al.*, 2017; Echa, 2020).

For reactive extraction, the ratio of 1:1 was used between 5 wt% D2EHPA in paraffin oil and medium. For optimal extraction, 2,4-toluenediamine has to be protonated, while D2EHPA not. Thus, reactive extraction should occur at an equilibrium pH above the pKa of D2EHPA (2.7) and below the pKa of 2,4-toluenediamine (5.3) to enable the formation of ionic bonds of the deprotonated extractant with the protonated amine. Physical extraction of deprotonated 2,4-toluenediamine at higher pH values is also possible, but this is far less efficient. Also, high pH values cause more partitioning of D2EHPA into the aqueous phase (Biswas *et al.*, 2000).

Table 8 Influence of pH on the efficiency of 2,4-toluenediamine (2,4-TDA) extraction, from a mixture with adipic acid, 1,4-butanediol and ethylene glycol (60 mM carbon equivalent of each), with 5% D2EHPA in paraffin oil. Errors depict the deviation from the mean (n = 2) (Utomo *et al.*, 2020).

Extraction pH	**Equilibrium pH**	**Separation efficiency of 2,4-TDA***	**Separation efficiency***			**Substrate utilization efficiency****		
			AA	**1,4-BDO**	**EG**	**AA**	**1,4-BDO**	**EG**
4	3.25 ± 0.02	93% ± 0.1%	0.0% ± 0.0%	1.2% ± 1.0%	0.3% ± 0.0%	100% ± 0%	100% ± 0%	100% ± 0%
5	3.93 ± 0.01	94% ± 0.2%	0.3% ± 0.0%	1.4% ± 1.0%	0.0% ± 0.0%	100% ± 0%	52% ± 1%	80% ± 3%
6	4.73 ± 0.01	94% ± 0.3%	1.4% ± 0.1%	1.7% ± 0.8%	0.0% ± 0.0%	9% ± 2%	62% ± 1%	81% ± 2%
7	5.30 ± 0.01	94% ± 0.3%	1.4% ± 0.2%	2.2% ± 0.1%	0.0% ± 0.0%	2% ± 0%	0% ± 0%	5% ± 0%

*: Efficiency of 100% means that the monomer was completely extracted to organic phase

** Growth of a mixed culture of 4 *Pseudomonas* strains after extraction in 25 ml medium. Efficiency of 100% means that the monomer was completely consumed after 24 h.

After the extraction at different pH values, the aqueous phase's pH was adjusted to 6.7, and mineral salts, buffer, and ammonium sulfate were added as a medium for *Pseudomonas* (Hartmans *et al.*, 1989). Extraction at pH 4 enabled maximum separation efficiency of 2,4-toluenediamine with partition coefficient (5% D2EHPA in paraffin oil/ medium) of 13.29 ± 0.01, and full utilization of the three remaining plastic monomers in 24 h (table 8) by the subsequent mixed culture. Higher starting pH values also enabled efficient extraction, but here, the mixed culture's growth was inhibited more strongly, likely due to higher partitioning of D2EHPA into the aqueous phase (Bednarz *et al.*, 2017). This result is supported by initial biocompatibility tests with either pure D2EHPA or paraffin oil alone. Pure D2EHPA inhibits the mixed culture's growth while paraffin oil alone did not.

3.1.3 Growth of a defined mixed culture on extracted mock PU hydrolysate

The efficient extraction of 2,4-toluenediamine allows for the omission of *P. capeferrum* TDA1, thereby reducing the defined mixed culture to three dedicated *P. putida* KT2440 derivatives (see figure 10). To assess monomer utilization from a model polyester PU, we used a synthetic mixture derived from a manufacturing formulation. Poly (1,4-butylene ethylene adipate) with an average molecular weight of 2630 (range: 950-4310) was used as a representative of a commercially available polyester (Songwon, 2017).

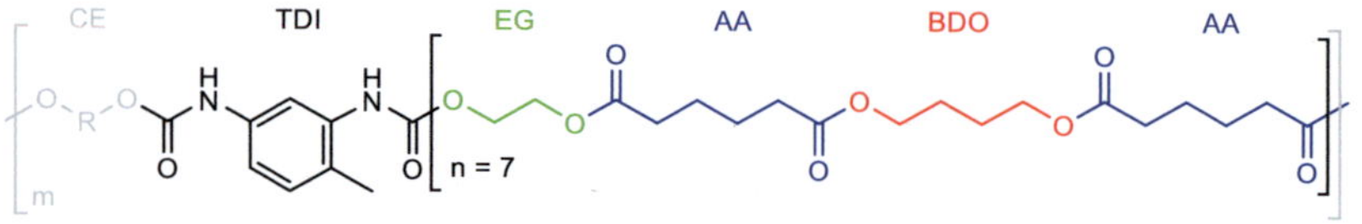

Figure 8 Model PU structure with AA: 1,4-BDO: EG: TDI molar ratio of 14:7:7:1. CE: chain extender, TDI: 2,4-toluene diisocyanate, EG: ethylene glycol, AA: adipic acid, 1,4-BDO: 1,4-butanediol (Utomo *et al.*, 2020)

By assuming an equal ratio of both diols, this average molecular weight would contain an estimated seven repeating units of EG-AA-1,4-BDO-AA. Further assuming a 1:1 ratio of polyester to 2,4-toluene diisocyanate and disregarding the chain extender for the sake of simplicity, the molar ratio of monomers resulting from full hydrolysis of this theoretical PU (figure 8) would be 14:7:7:1 (AA: 1,4-BDO: EG: 2,4-TDA). If hydrolysis starts with 24 g L^{-1} PU, the resulting hydrolysate would contain approximately 121 mM adipic acid, 60 mM ethylene glycol, 60 mM 1,4-butanediol, and 8.5 mM 2,4-toluenediamine.

Since low extraction pH for 2,4-toluenediamine allowed full utilization of the remaining monomers by the microbial mixture, a more detailed investigation of the extraction pH was conducted with this mock PU hydrolysate. Therefore, reactive extraction of 2,4-toluenediamine was performed at pH 3.5, 4.0, and pH 4.5, and the supernatant was subsequently neutralized, supplemented with mineral medium components, and inoculated with the mixed culture of three *P. putida* KT2440 strains (table 9).

Table 9 The separation efficiencies and growth of the defined mixed culture on mock PU hydrolysate extracted with 5% D2EHPA in paraffin oil. Errors depict the deviation from the mean (n = 2) (Utomo *et al.*, 2020).

Extraction pH	Equilibrium pH	Separation efficiency of 2,4-TDA*	Separation efficiency* AA	1,4-BDO	EG	µ max**	Final OD**
3.5	3.29 ± 0.02	93% ± 0.3%	1.0% ± 0.4%	1.2% ± 0.7%	0.0% ± 0.0%	0.18 ± 0.00	1.81 ± 0.21
4.0	3.82 ± 0.02	94% ± 0.6%	1.5% ± 0.2%	0.8% ± 0.3%	0.8% ± 0.6%	0.13 ± 0.00	1.95 ± 0.06
4.5	4.41 ± 0.01	94% ± 0.2%	1.5% ± 0.0%	0.6% ± 0.0%	0.0% ± 0.0%	0.11 ± 0.00	0.99 ± 0.33

*: Efficiency of 100% means that the monomer was completely extracted to organic phase.

**: Growth of a mixed culture of 3 *P. putida* KT2440 strains in 2 ml medium with 24-square-well Growth Profiler and System Duetz plates for 210 hours.

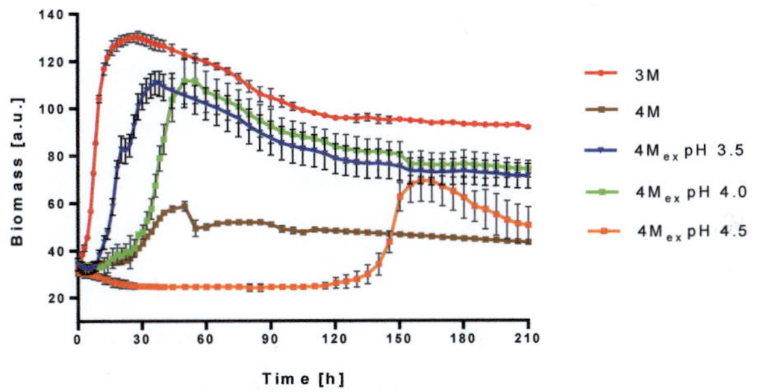

Figure 9 Growth of a mixed culture of 3 *P. putida* KT2440 strains in MSM with mock PU hydrolysate as carbon source, with ($4M_{ex}$) and without (4M) extraction of 2,4-toluenediamine with a 1:1 ratio of 5 wt% D2EHPA in paraffin oil at the indicated pH. Mock hydrolysate without 2,4-toluenediamine consisting only three monomers (3M) was included as a positive control. The 3 strains were pre-cultured individually then mixed with proportional OD as a defined mixed culture in 2 ml medium in 24-square-well Growth Profiler plates. Errors depict the deviation of the mean (n = 2) (Utomo *et al.*, 2020).

The extraction of 2,4-toluenediamine at pH 3.5 ($4M_{ex}$ pH 3.5) led to a shorter lag phase and a higher growth rate of 0.18 h^{-1} than $4M_{ex}$ pH 4.0, which enabled a growth rate of 0.13 h^{-1} (figure 9). On $4M_{ex}$ pH 4.5, a long lag phase occurred, but eventually, growth could be observed after 130 hours. From this experiment, $4M_{ex}$ pH 3.5 proved optimal, and it significantly enhanced the growth of the mixed culture compared to the original 4M containing 2,4-toluenediamine (figure 9). However, growth was still slower compared to the mock hydrolysate without 2,4-toluenediamine consisting only three monomers (3M) as a control (μ = 0.21 ± 0.006 h^{-1}), indicating that the extractant is not 100% biocompatible. A toxicity effect of 2,4-toluenediamine is unlikely because all three pH values led to the same extraction efficiency (table 9). This toxicity of the extractant might be avoided by the use of more solvent-tolerant *Pseudomonas* chassis strains and possibly amended by laboratory evolution to enhance tolerance (Heipieper *et al.*, 2007; Wynands *et al.*, 2019). Alternatively, a more biocompatible extractant could be used.

3.1.4 PU monomers as a substrate for rhamnolipid production

After the establishment of the extraction and microbial utilization of a mock PU hydrolysate, the mixed culture consisting of *P. putida* KT2440 derivatives was also used to produce mono-rhamnolipids as a proof-of-concept of plastics monomer upcycling (figure 10). *P. putida* KT2440 was previously engineered for mono-rhamnolipid production from glucose by expressing two genes from *Pseudomonas aeruginosa,* encoding RhlA and RhlB (Tiso *et al.*, 2016). RhlA converts two ACP-activated OH-fatty acids into 3-(3-hydroxyalkanoyloxy) alkanoic acid (HAA), while the rhamnosyltransferase RhlB subsequently adds a rhamnose to the HAAs using dTDP L-rhamnose as a substrate (Zhu and Rock, 2008). These genes were transformed into all three strains of the defined mixed culture on plasmid pPS05 (Tiso *et al.*, 2016 and figure 10). The simple implementation of a recombinant production pathway was possible, as the defined mixed culture was purposefully based on a single reference strain, *P. putida* KT2440.

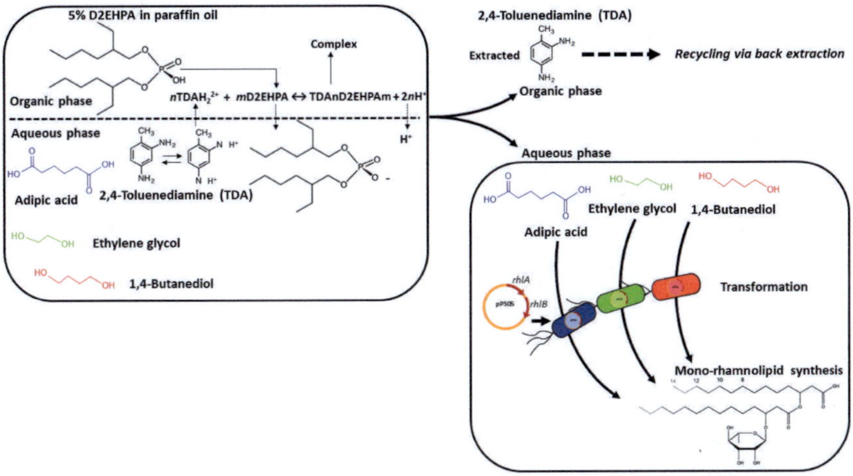

Figure 10 Workflow from PU hydrolysate to glycolipid synthesis. **Left:** Reactive extraction by D2EHPA in paraffin oil of toxic 2,4-toluenediamine. **Right:** The extracted 2,4-toluenediamine could further be used for diisocyanate synthesis after back extraction. Growth and rhamnolipid production of a mixed culture of three engineered or evolved *P. putida* KT2440 substrate specialist strains transformed with plasmid pPS05 on the remaining PU monomers in aqueous phase (Utomo *et al.*, 2020).

Each single strain, cultivated on their preferred PU monomer as sole C-source, was able to produce mono-rhamnolipids. *P. putida* B10.1 grown on 15 mM 1,4-butanediol achieved the highest biomass (figure 11) as well as the highest titer and yield of mono-rhamnolipids among the three strains (table 10).

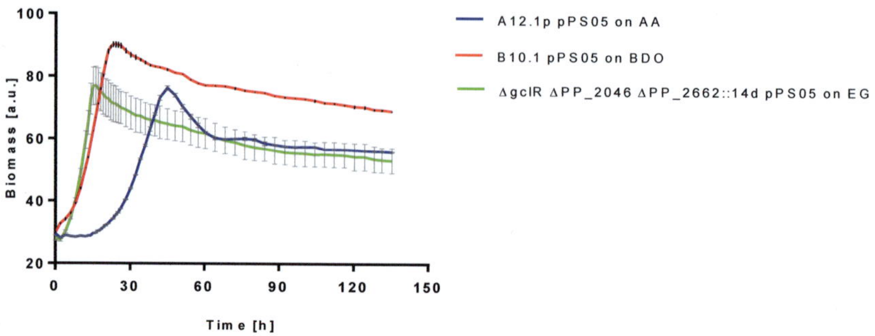

Figure 11 Growth of 3 *P. putida* KT2440 strains (A12.1p, B10.1, and Δ*gclR* ΔPP_2046 ΔPP_2662::14d) harboring plasmid pPS05 for mono-rhamnolipid production in MSM supplemented with specific monomer as their sole carbon sources (AA: adipic acid; 1,4-BDO: 1,4-butanediol; EG: ethylene glycol. Growth was accessed by growth profiler in 2 ml medium in 24-square-well plates. Errors depict the deviation of the mean (n = 2) (Utomo *et al.*, 2020)

The mixed culture was able to fully consume the plastic monomer mixture, but mono-rhamnolipids production was lower than the production of the single strains on a single monomer (table 10). The extraction of 2,4-toluenediamine somewhat lowered the mixed culture's growth and productivity compared to the mock hydrolysate without 2,4-toluenediamine consisting only three monomers (3M) as a control, but it clearly outperformed the non-extracted 4M mixture (figure 12). The maximum consumption rates (q_s) of adipic acid, 1,4-butanediol, and ethylene glycol in the $4M_{ex}$ mixture were 0.25 ± 0.09, 0.75 ± 0.03, and 0.62 ± 0.16 mmol g_{cdw}^{-1} h^{-1} respectively.

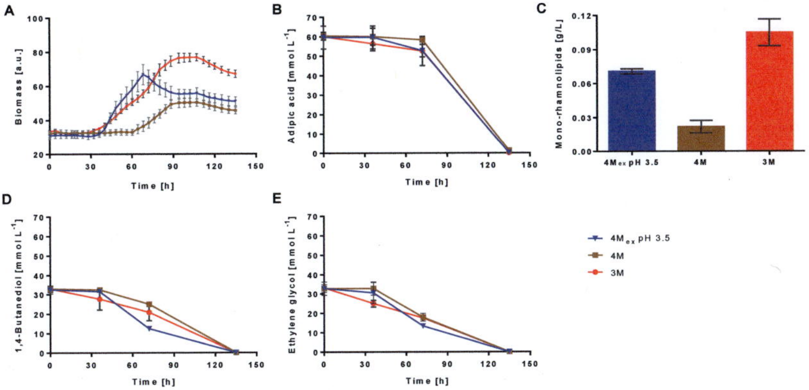

Figure 12 Growth of a mixed culture of 3 *P. putida* KT2440 strains, the utilization of the mock PU hydrolysate for mono-rhamnolipid production, and the concentration of mono-rhamnolipids were observed. **A**: The growth of mixed culture in MSM with mock PU hydrolysate as carbon source, with ($4M_{ex}$) and without (4M) extraction of 2,4-toluenediamine with a 1:1 ratio of 5 wt% D2EHPA in paraffin oil at pH 3.5. Mock hydrolysate without 2,4-toluenediamine consisting only three monomers (3M) was included as a positive control. The utilization in each monomer of the mock PU hydrolysate: **B**: adipic acid (60.5 mmol L^{-1}), **C**: 1,4 butanediol (30 mmol L^{-1}), and **D**: ethylene glycol (30 mmol L^{-1}). **E**: The final concentration of mono-rhamnolipids in MSM with 3M, 4M, and $4M_{ex}$ as carbon source. The 3 strains were pre-cultured individually then mixed with proportional OD as a defined mixed culture in 2 ml medium in 24-square-well Growth Profiler plates. Errors depict the deviation of the mean (n = 2) (Utomo *et al.*, 2020).

For comparison, the yields of rhamnolipids from the plastics monomers 1,4-BDO and EG were higher than those reported on the industrial substrates, i.e., glycerol or molasses (Ochsner *et al.*, 1995; Onbasli and Aslim, 2009). However, higher yields were reported using glucose as a carbon source (Tiso *et al.*, 2016). Regarding the envisaged circular bioeconomy, this study has successfully shown that plastic monomers could be utilized as novel substrates and biotechnologically used for rhamnolipid production.

Table 10 Rhamnolipid production from PU monomers and extracted mock PU hydrolysate compared to literature values (Utomo *et al.*, 2020)

Strain	Medium*	Substrate	Substrate [g/L]	CDW [g/L]	µ max** [h^{-1}]	Rhl*** [g/L]	P/S yield [g/g]	Ref.
P. aeruginosa PAO1	MS	Sunflower oil	250	16.3	0.13	39	0.156	(Müller *et al.*, 2010)
P. putida KT2440 pPS05	LB	Glucose	11	3.4	0.90	2.2	0.200	(Tiso *et al.*, 2016)
P. putida KT2442	GS	Glycerol	20	n.d	n.d	0.6	0.030	(Ochsner *et al.*, 1995)
P. putida B12	DM	Molasses	50	n.d	n.d	0.52	0.010	(Onbasli and Aslim, 2009)
P. putida KT2440 A12.1p pPS05	MS	AA	1.46	1.42 ± 0.15	0.14 ± 0.00	0.02 ± 0.01	0.014 ± 0.00	This study
P. putida KT2440 B10.1 pPS05	MS	1,4-BDO	1.35	2.22 ± 0.00	0.16 ± 0.01	0.13 ± 0.01	0.088 ± 0.00	This study
P.putida KT2440 Δ*gclR* ΔPP_2046 ΔPP_2662::14d	MS	EG	1.86	1.56 ± 0.13	0.29 ± 0.01	0.07 ± 0.00	0.038 ± 0.00	This study
Defined mixed culture + pPS05	MS	3M	13.3	1.65 ± 0.00	0.09 ± 0.02	0.10 ± 0.01	0.008 ± 0.00	This study
	MS	4M	13.3	0.55 ± 0.00	0.06 ± 0.01	0.02 ± 0.01	0.002 ± 0.00	This study
	MS	$4M_{ex}$ pH 3.5	13.3	1.38 ± 0.12	0.12 ± 0.00	0.07 ± 0.00	0.005 ± 0.00	This study

*: MS: Mineral salts; LB: Lysogeny broth; GS: Glycerol; DM: Diluted molasses; **: n.d: not determined; ***: Rhl: Rhamnolipids.

3.2 Simultaneous utilization of PET monomers by *Pseudomonas capeferrum* TDA1 for rhamnolipid synthesis

Romualdus Nugraha Catur Utomo[1], Katrin Meier[1], Sebastian Kobbing[1], Tobias B. Alter[1], Till Tiso[1], Birgitta Ebert[1], Tanja Narancic[2], Kevin O'Connor[2], Herman J. Heipieper[3], Christian Eberlein[3], Nick Wierckx[4], Lars M. Blank[1]

[1]Institute of Applied Microbiology-iAMB, Aachen Biology and Biotechnology-ABBt, RWTH Aachen University, Aachen, Germany

[2]UCD Earth Institute and School of Biomolecular and Biomedical Science, University College Dublin, Dublin, Ireland

[3]Department of Environmental Biotechnology, Helmholtz Centre for Environmental Research - UFZ, Leipzig, Germany

[4]Forschungszentrum Jülich Institute of Bio- and Geosciences IBG-1: Biotechnology, Jülich, Germany.

Manuscript in preparation

Contributions:

1. This chapter was written by Romualdus Nugraha Catur Utomo and was revised by Lars M Blank. Romualdus Nugraha Catur Utomo performed ALE experiment, transformation of *tph* operon-harboring pBT'T, Tn7 integration, pH optimization for rhamnolipids production, and fed-batch culture in the shake flasks for improvement of rhamnolipids production.
2. Tanja Narancic provided plasmid pBT'T encoding the *tph* operon for terephthalic acid utilization
3. Till Tiso provided pPS05 and pSK02 for mono-rhamnolipid synthesis
4. Sebastian Köbbing constructed the *tph* operon with Tn7 for genomic integration
5. Tobias Alter developed the mini bioreactor system.
6. Katrin Meier conducted experiments on mini bioreactors and confirmation test on growth profiler as a part of her master's thesis (Meier, 2020).

3.2.1 Adaptive laboratory evolution of *P. capeferrum* TDA1

P. capeferrum TDA1 has interesting capacities towards plastic monomer utilization (Cárdenas Espinosa *et al.*, 2020), and hence its portfolio of monomer degradation pathways should be extended to ethylene glycol and terephthalic acid, the monomers of PET. *P. capeferrum* TDA1 was initially grown in a mineral salt medium (MSM) with 10 mM adipic acid as a sole carbon source. Since the strain natively grows on that substrate, this condition was chosen as pre-culture. It is also shown from a previous study that this strain also can utilize 1,4-butanediol at a moderate level. Therefore, 1st main-culture (10 mM adipic acid) and 2nd main culture (2 mM adipic acid + 15 mM 1,4-butanediol) were intended to evolve the strain growing better on 1,4-butanediol as its second plastic monomer substrate.(Utomo *et al.*, 2020) The 3rd main culture was a critical step to trigger the strain growing on ethylene glycol by adding 5 mM glyoxylic acid as an intermediate into 2 mM adipic acid and 2 mM 1,4-butanediol (Mückschel *et al.*, 2012). The glyoxylate carboligase (Gcl) converts two glyoxylate molecules into tartronate semialdehyde and CO_2. This route is considered the possible pathway to utilize glyoxylate as the sole carbon source (Li *et al.*, 2019).

Furthermore, the 4th main culture consisted of 1 mM adipic acid, 1 mM 1,4-butanediol, 2 mM glyoxylic acid, and 20 mM ethylene glycol. Growth was slow while reducing ethylene glycol in the following cultures 5th and 6th to 8 mM resulted in a faster adaptation of strain TDA. Finally, the evolved strain could grow on 30 mM ethylene glycol as sole carbon source. For the confirmation, the evolved strain was subsequently tested on a single monomer of 10 mM adipic acid, 15 mM butanediol, and 30 mM ethylene glycol in MSM with observed final OD_{600} of 1.3, 1.9, and 1.8, respectively (Figure 13 B).

Table 11 Adaptive laboratory evolution (ALE) plan on various media

Step	Medium	Incubation (day)
Pre-culture	10 mM Adipic acid	1
1st Main culture	2 mM Adipic acid + 15 mM 1,4-Butanediol	2
2nd Main culture	15 mM 1,4-Butanediol	3
3rd Main culture	2 mM Adipic acid + 2 mM 1,4-Butanediol + 5 mM Glyoxylic acid	1
4th Main culture	1 mM Adipic acid + 1 mM 1,4-Butanediol + 2 mM Glyoxylic acid + 20 mM Ethylene glycol	1
5th Main culture	1 mM Adipic acid + 3 mM 1,4-Butanediol + 1.6 mM Glyoxylic acid + 8 mM Ethylene glycol	3
6th Main culture	1 mM Adipic acid + 3 mM 1,4-Butanediol + 1.6 mM Glyoxylic acid + 8 mM Ethylene glycol	3

7th Main culture	30 mM Ethylene glycol	5
Total ALE incubation (day)	:	19

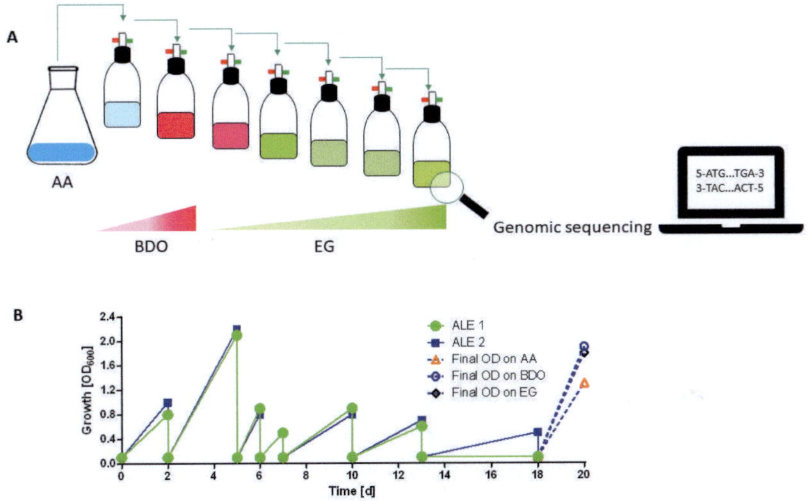

Figure 13 Adaptive laboratory evolution of *P. capeferrum* TDA1 on plastic monomers. **A**: Sketch of experimental design. **B**: ALE experiment (see table 11 for media details) and the successful evolved strain on ethylene glycol (ALE 1) was subsequently tested on single monomer EG, AA, and BDO.

The *gntR* family of transcription regulators was firstly described in 1991 and named after the gluconate repressor in *Bacillus subtilis.* This *gntR* family is considered a large group of proteins regulating various biological processes in bacteria (Suvorova *et al.*, 2015). From the alignment of 24 *gntR* family transcriptional regulators in KT2440 towards *P. capeferrum* TDA1, 17 *gntRs* were found in *P. capeferrum* TDA1 with ≥ 80 % of matches.

Table 12 List of *gntR* family transcriptional regulator in *P. capeferrum* TDA1 based on the alignment against *P. putida* KT2440

P. putida KT2440		Location in *P. capeferrum* TDA1 genome		Matches (%)
Found in strain	Length	Position in genome	Length	
PP_0486	1425	590474-589061 C	1408	82
PP_ 0620	765	744918-744154 C	765	91
PP_ 0969	651	1079972-1080577	606	84
PP_1727	714	1407897-1408603	707	86
PP_2253	669	3922045-3922681	637	85
PP_2254	699	3921308 -3921999	686	89
PP_2333	717	3872509-3871377 C	683	91
PP_2542	1425	2126432-2125014 C	1419	89
PP_2642	1491	3490515-3489052 C	1458	80
PP_ 3544	1400	2461067-2459668 C	1400	88
PP_ 3603	765	2380502-2381265	765	91
PP_3649	753	2332227-2331463 C	753	89
PP_3750	1434	3685803-3687236	1434	85
PP_4197	1593	4069998-4068475 C	1498	85
PP_4277	660	4148245-4147595 C	651	87
PP_4283	756	4156291-4155536 C	756	91
PP_5342	1530	5895212-5893859 C	1353	80

From a previous study, *gclR* (PP_4283) in KT2440, one of the *gntR* family transcriptional regulators, is a transcriptional repressor of the operon encoding glyoxylate carboligase (Gcl) pathway (Li *et al.*, 2019). From resequencing analysis and genome variant detection, it is observed that there are two mutations related to the *gntR* family transcriptional regulators, which are identical to PP_2254 and PP_4283 in *P. putida* KT2440. Since there is still no consensus to name all genes in *P. capeferrum* TDA1, homologues genes to PP_2254 and PP_4283 are called PTDA1_2254 and PTDA1_4283 in this study. Two deletions occurred in PTDA1_2254 in the position of 3921563 (GC->G) and the position of 3921566

(TCGGCCAGCCTGGCCGATGTGCTGGAGCTGCGCTACACCCTCG ->T) of the genome. On the other side, mutations occurred in the intergenic region of PTDA1_4283 upstream in the position of 4156296 (gTg/gAg) and in the position of 4156346 (AGGGTGGCTGGCAAGTCCT->A). Adaptive mutation in the intergenic region received less attention than in the coding region. However, it was shown in the study from Khademi et al. (2019) that the intergenic mutations were more frequently located upstream of known transcriptional start sites and could facilitate essential genes as targets of evolution.

The transcriptional regulation of both genes was investigated using reverse transcription qualitative PCR (RT-qPCR). As a sole carbon source in MSM, glucose was used to grow the wild type and ALE TDA1. Furthermore, total RNA was isolated as a template for RT-qPCR, then analyzed using three different methods, namely Livak, Pfaffl, and multiple housekeeping genes (HKG) methods. Since *rho* and *rpoD* genes were reported as the most suitable reference genes in many bacteria, including *Pseudomonads*, these two genes were also applied in this study for accurate RT-qPCR data normalization on wild type and ALE *P. capeferrum* TDA1 (Gomes *et al.*, 2018; Bai *et al.*, 2020).

As described before, the Livak method compares the C_t (cycle threshold) values from different RNA samples with C_t values of HKG for normalization. This method does not consider differences in primer efficiencies (Livak and Schmittgen, 2001). The analysis focuses on the gene expression fold change of two *gntR* in the wild type and ALE *P. capeferrum* TDA1. The result showed high expression fold change of both *gntR* only in the wild type. The highest expression was from *gntR* PTDA1_2254 with HKG *rpoD*, which reached 2021.30 expression fold change. From the Pfaffl method, the primer efficiencies were also involved in the analysis. The result showed higher expression fold change of both *gntR* on wild type with HKG *rho*. From multiple HKG analyses, the strong *gntR* expression was only obtained from wild type, while ALE was not. The results show that transcription of both genes was absent in ALE TDA1, which is in good agreement with the observed growth on EG.

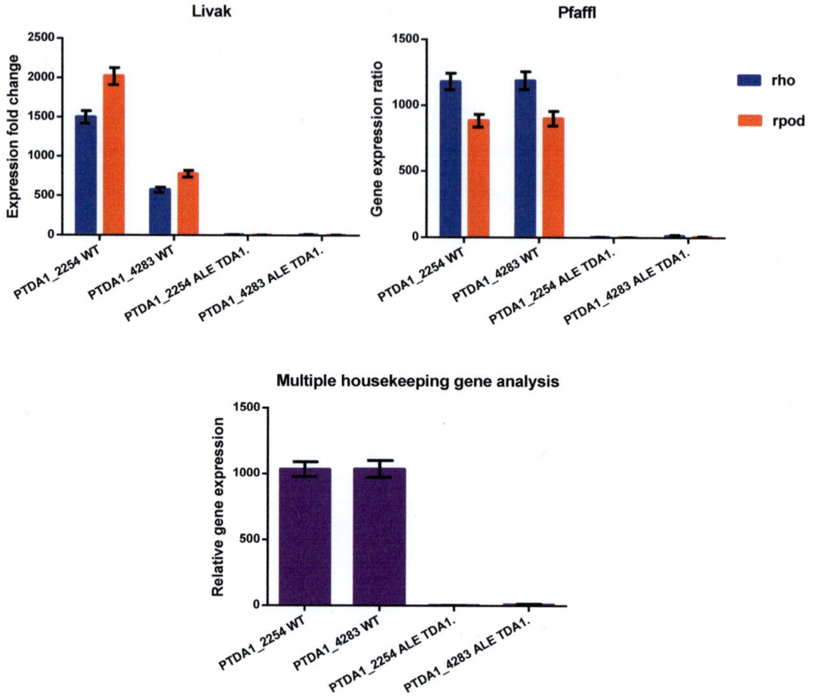

Figure 14 Analysis of transcription of the *gntR* genes PTDA1_2254 and PTDA1_4283 during growth on MSM medium with glucose a sole carbon source. The result of RT-qPCR was analyzed by using Livak, Pfafll, and multiple housekeeping gene analysis with *rho* and *rpoD* as reference genes for normalization.

3.2.2 Terephthalic acid utilization

Pseudomonas umsongensis GO16, isolated from a soil sample exposed to PET granules at a PET bottle processing plant, can utilize terephthalic acid as a sole carbon source and is known as PHA producer (Kenny *et al.*, 2008). The strain harbors the genetic inventory as previously described TA utilizers like *Comamonas pcaRtphCA2A3BA1* (Narancic *et al.*, 2021). Here, the *tph* operon was cloned and either transformed via a plasmid or integrated via the Tn7 system into *P. capeferrum* TDA1 to utilize terephthalic acid.

Plasmid-based transcription allows high enzyme activities, as the plasmid might exist in high copy numbers. The limitations of plasmids are well-known, covering stability and antibiotic use. Therefore, the genetic scheme with Tn7 mini transposons targeting the *attTn7* sequence would be a great alternative to implant genes inside the genome for specific purposes in bioremediation, biocatalysis, or biosensing in the environment (De Las Heras *et al.*, 2008). This single *att*Tn*7* site is close to *glmS*, which encodes the essential enzyme glucosamine-6-phosphate synthetase (Y. J. Choi *et al.*, 2006). Using a Tn7 derivative carrying the *tph* operon of *P. umsongensis* GO16, strain TDA1 was modified. From de novo assembly of whole-genome sequencing of the *tph* operon-integrated *P. capeferrum* TDA1 and BLAST analysis against the *tph* operon sequence by Eurofins Genomics (Ebersberg, Germany), it has been confirmed that the *tph* operon was successfully integrated into the downstream of the essential glmS gene (Choi *et al.*, 2005), which is located in the position of 5971591-5969756 C of *P. capeferrum* TDA1 genome (see figure S3 in Appendix).

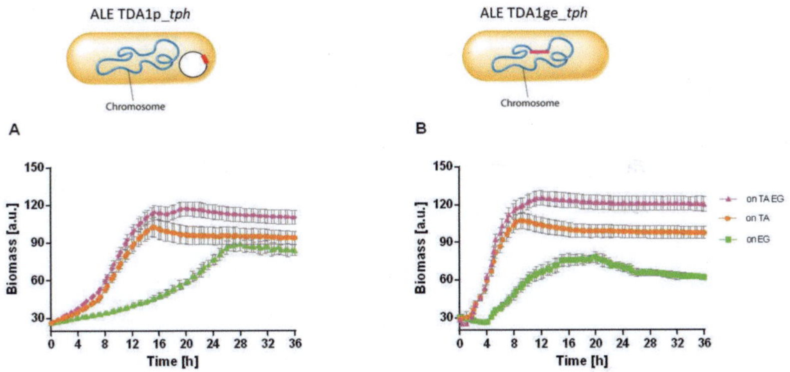

Figure 15 Evolved strain of *P. capeferrum* TDA1 harboring the *tph* operon via plasmid (ALE TDA1p_*tph*) (**A**) and via genomic integration (ALE TDA1ge_*tph*) (**B**). Growth was accessed using the growth profiler in 2 ml medium in 24-square-well plates. Errors depict the deviation of the mean (n = 2).

Table 13 Growth rate comparison between ALE TDA1p_*tph* and ALE TDA1ge_*tph*

Strain	Substrate	μ max [h^{-1}]
ALE TDA1p_*tph*	Ethylene glycol	0.09 ± 0.00
	Terephthalic acid	0.21 ± 0.00
	Ethylene glycol + Terephthalic acid	0.22 ± 0.00
ALE TDA1ge_*tph*	Ethylene glycol	0.21 ± 0.01
	Terephthalic acid	0.38 ± 0.01
	Ethylene glycol + Terephthalic acid	0.46 ± 0.02

Overall, in all conditions, ALE TDA1ge_*tph* had a higher maximum growth rate than ALE TDA1p_*tph*. It is an excellent indication that the integrated *tph* operon expressed well. The system would be perfectly suitable for large-scale applications, such as fermenters or in situ applications in the context of PET monomer utilization from plastic waste to value-added products. Without antibiotics as a selective marker, the strain would be environment-friendly, provide more stable single-copy integration, and not require selective pressure (Choi *et al.*, 2005; De Las Heras *et al.*, 2008; Crépin *et al.*, 2012).

3.2.3 Further evolution in mini bioreactor with sequential and turbidostatic feeding system

The evolved strain, ALE TDA1ge_*tph*, could grow on ethylene glycol through ALE and terephthalic acid through the Tn7 integration system. However, further evolution was needed to enable the strain to grow better on a mixture of both PET monomers. Therefore, the investigation was conducted using a small-scale fermenter to evolve the strain under a certain evolutionary pressure by serially diluting the culture with a sequential and turbidostatic system (Toprak *et al.*, 2012; Yang, 2018).

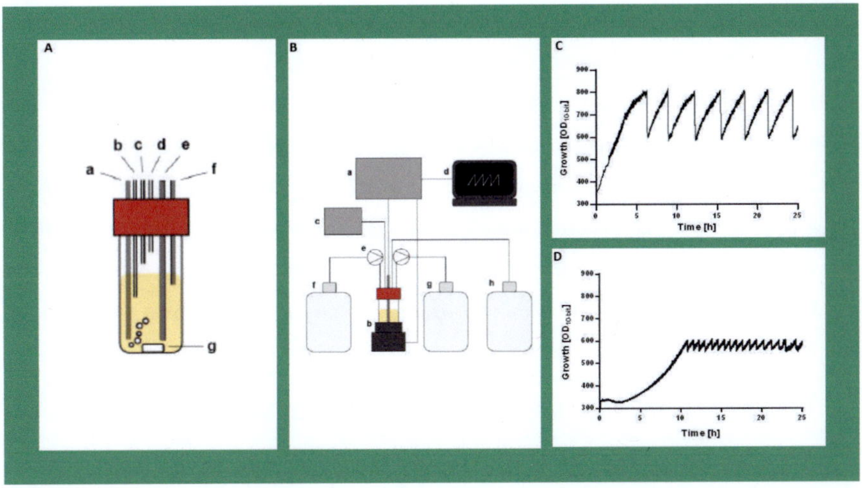

Figure 16 Workflow of further evolution by using a mini bioreactor. A. Fermenter setup: Fermenter vial with tubes for air inlet (a), bleed withdrawal (b), air outlet (c), feeding (d), temperature sensor (e), and sampling (f). **B.** Fermentation system with an electromechanical relay interface device (a). The relay interface includes a temperature sensor that is connected to the inlet of the fermentation vial. The vial sits in a tube holder, including an infrared-turbidity sensor, a magnetic stirrer, and heating at the bottom (b). Aeration is assured by an air pump (c). The feeding program is started over the graphic user interface (d), which allows the pumps (e) to add fresh medium from the feed bottle (f) into the fermenter and subsequent withdrawal of cell culture from the fermenter into the bleed bottle (g). Foam and pressure can escape through tubing into the air outlet bottle (h). **C.** Dilution scheme of sequential feeding system. **D.** Dilution scheme of turbidostatic feeding system (Meier, 2020)

3.2.3.1 Sequential feeding system

In the beginning, sequential feeding was established for obtaining an enhanced growth rate as the indicator of evolution. After a set threshold OD was reached in a sequential batch system, 10 mL of broth were automatically pumped out of the fermenter into the bleed vessel. The exact amount of fresh medium was pumped into the fermenter after a certain time delay. Subsequently, a lower OD was measured, and the process was repeated. In this system, the culture is constantly kept in the exponential phase.

During this experiment, the growth rate on 20 mM terephthalic acid in MSM increased from 0.14 h^{-1} to 0.16 h^{-1} between 25 and 125 hours of cultivation. Subsequently, the feed was switched to 30 mM ethylene glycol in mineral salt medium (MSM), and a definite growth rate decrease was observed to a value of 0.04 h^{-1}. First, a further decrease is seen until 175 h to 0.02 h^{-1}, but after that, an increase is noticed to 0.06 h^{-1} at 210 h. The growth rate jump between 175 h and 210 indicates mutations through selection pressure towards ethylene glycol utilization. After the feed was changed from 20 mM terephthalic acid to 30 mM ethylene glycol in MSM, the growth rate dropped, and much smaller growth rates were noticed, indicating that terephthalic acid is the preferred carbon source (Meier, 2020).

A sample was taken from fermenter 1 at the fermentation end, and the sample was streaked out on an LB agar plate. Single colonies were isolated from screening and selecting the clone with the highest biomass formation in System-Duetz by means of their end-OD600 after growing on the mixture of ethylene glycol and terephthalic acid. As a result, the new variant ALE TDA1ge_*tph* F1.1 was obtained, further grown on each monomer and a monomer mixture compared to the variant ALE TDA1ge_*tph* using the growth profiler. The additional mutation also occurred in this strain upstream of PTDA1_2253 (*gntR* family of transcription regulators) in the position of 3922024 (gCg/ gTg) of the genome.

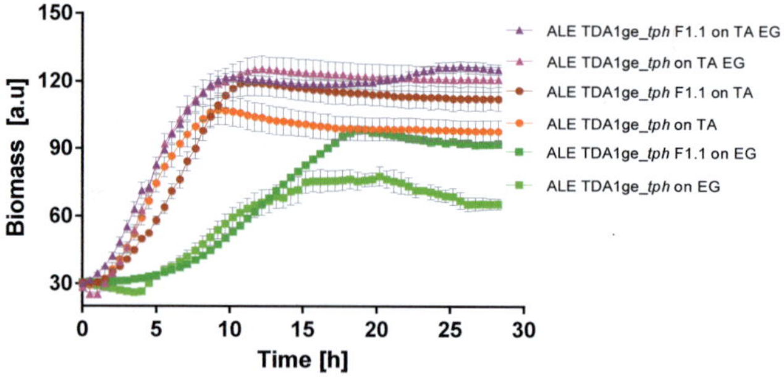

Figure 17 Growth of *ALE TDA1ge_tph* and ALE TDA1ge_*tph* F1.1 on single and mixed PET monomers. Growth was accessed using the growth profiler in 2 ml medium in 24-square-well plates. Errors depict the deviation of the mean (n = 2) (Meier, 2020).

Table 14 Growth rate comparison between *ALE TDA1ge_tph* and *ALE TDA1ge_tph* F1.1 (Meier, 2020)

Strain	Substrate	μ max [h^{-1}]
ALE TDA1ge_*tph*	Ethylene glycol	0.21 ± 0.01
	Terephthalic acid	0.38 ± 0.00
	Ethylene glycol + Terephthalic acid	0.46 ± 0.02
ALE TDA1ge_*tph* F1.1	Ethylene glycol	0.15 ± 0.01
	Terephthalic acid	0.34 ± 0.01
	Ethylene glycol + Terephthalic acid	0.45 ± 0.01

From further ALE in the mini bioreactor, the ALE TDA1ge_*tph* F1.1 strain resulted in better biomass than ALE TDA1ge_*tph* on single monomer ethylene glycol or terephthalic acid as a sole carbon source. However, both strains had similar biomass on a mixture of ethylene glycol and terephthalic acid. Even though the maximum growth rate was slightly decreasing from the previous ALE, ALE TDA1ge_*tph* F1.1 had a higher substrate uptake rate (1.90 ± 0.07 mmol g_{cdw}^{-1} h^{-1}) and shorter lag phase (2 h) with ethylene glycol as the substrate. Based on the result, ALE TDA1ge_*tph* F1.1 was selected for a further purpose, namely rhamnolipid production and genome sequencing.

3.2.3.2 Turbidostatic feeding system

For further improvement, a turbidostatic feeding system was chosen as an alternative way to improve the growth rate on ethylene glycol as a sole carbon source. While the chemostat experiment has a fixed volume and dilution rate resulting in a fixed growth rate, turbidostatic cultivation is conducted by adjusting the flow rate to make the turbidity constant, and cells are kept at their maximum growth rate.(Watson, 1972; Toprak *et al.*, 2012) During this experiment, the maximum growth rate only reached around 0.08 h^{-1}, but still higher than the previous improvement with sequential feeding. Several series of cultivation on PET monomers using a growth profiler were conducted to see the phenotypes of the strains from sequential and turbidostatic systems.

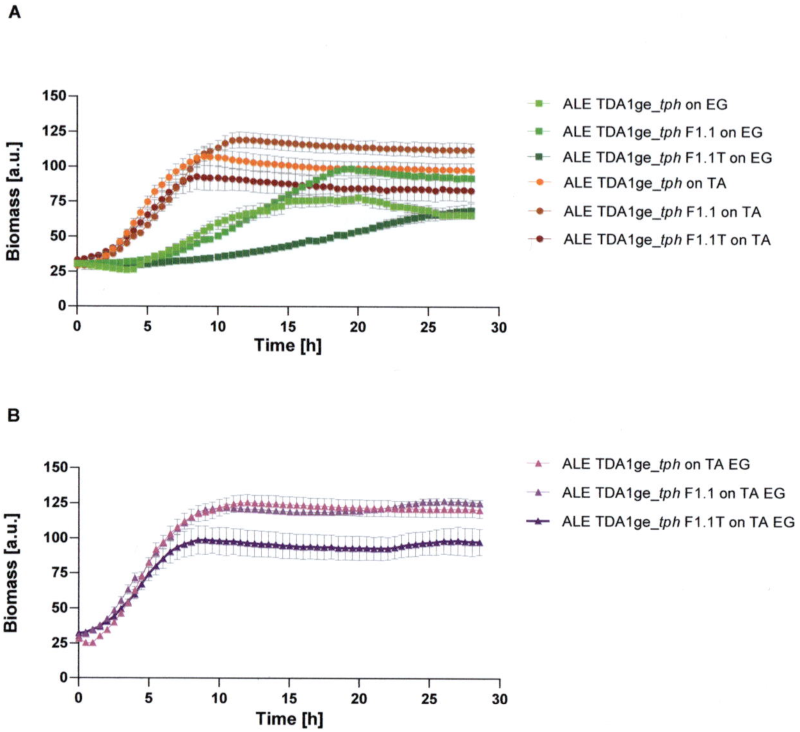

Figure 18 Growth of strain ALE TDA1 evolved in sequential feeding system **(**F1.1) and further evolved in turbidostatic feeding system (F1.1T) on single (**A**) and mixed PET monomers (**B**). Growth was accessed by using the growth profiler in 2 ml medium in 24-square-well plates. Errors depict the deviation of the mean (n = 2) (Meier, 2020).

As depicted in figure 18, the further evolution by turbidostatic system gave no significant improvement than previous ALEs; even *tph* ALE TDA1ge (without ALE in mini bioreactor) outperformed in all conditions both on single monomer and mixed monomer of PET. ALE TDA1ge_*tph* F1.1 strain from a sequential feeding system was selected for the other purpose for rhamnolipid production from PET monomers.

Table 15 Growth rate comparison between ALE TDA1ge_tph, ALE TDA1ge_tph F1.1, and ALE TDA1ge_tph F1.1T (Meier, 2020)

Strain	Substrate	μ max [h^{-1}]
ALE TDA1ge_*tph*	Ethylene glycol	0.21 ± 0.01
	Terephthalic acid	0.38 ± 0.00
	Ethylene glycol + Terephthalic acid	0.46 ± 0.02
ALE TDA1ge_*tph* F1.1	Ethylene glycol	0.15 ± 0.01
	Terephthalic acid	0.34 ± 0.01
	Ethylene glycol + Terephthalic acid	0.45 ± 0.01
ALE TDA1ge_*tph* F1.1T	Ethylene glycol	0.09 ± 0.00
	Terephthalic acid	0.28 ± 0.02
	Ethylene glycol + Terephthalic acid	0.33 ± 0.02

3.2.4 pH optimization for rhamnolipid production

Since pH is crucial when cultivating acidic plastic monomers (adipic acid and terephthalic acid), different pH values were determined to find the best condition of PET monomer utilization and rhamnolipid production.

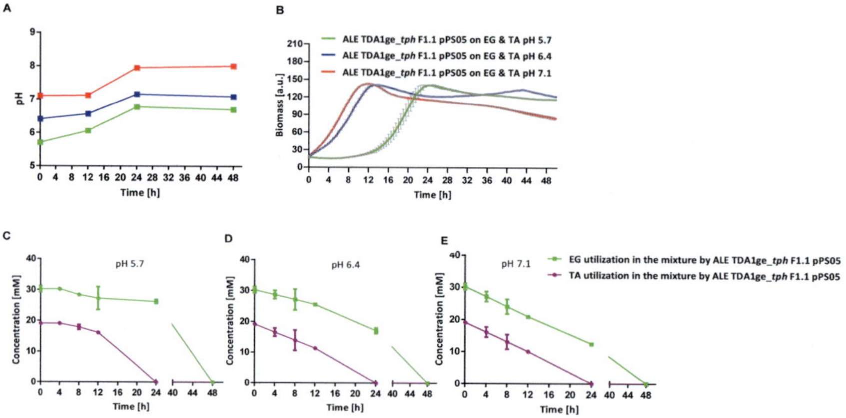

F

pH	Substrate (g/L)	μ Max (h^{-1})	Product (g/L)	Yield (P/S)
5.7		0.18±0.01	0.29±0.002	0.056±0.000
6.4	5.18	0.21±0.01	0.24±0.002	0.046±0.000
7.1		0.24±0.01	0.25±0.007	0.048±0.001

Figure 19 A: pH change over the cultivation time during rhamnolipids production corresponds to the growth (**B**) of *tph* ALE TDA1ge F1.1 pPS05, which was set with different initial pH. **C, D, E:** The utilization of ethylene glycol (EG) and terephthalic acid (TA) by ALE TDA1ge_*tph* F1.1 with initial pH of the cultures of 5.7, 6.4, and 7.1, respectively. **F:** the result of max. growth rate, rhamnolipid product, and yield of rhamnolipids production from mixed PET monomer. Growth was accessed by growth profiler in 2 ml medium in 24-square-well plates. Errors depict the deviation of the mean (n = 2).

This result showed that ALE TDA1_*tph* F1.1 was sensitive at lower pH affecting a more prolonged lag phase and resulting lowest growth rate (0.18 ± 0.01 h^{-1}) among three pH parameters, while the maximum consumption rates (q_s) of terephthalic acid and ethylene glycol was achieved at initial pH of 7.1 with the value of 1.33 ± 0.05 and 1.37 ± 0.12 mmol g_{cdw}^{-}

^{1} h^{-1}, respectively. Meanwhile, the concentration and yield in rhamnolipid production gave a similar result.

3.2.5 Fed-batch culture in the shake flasks for improvement in rhamnolipids production

Fed-batch cultivation was proposed to maintain the pH of culture growth and improve rhamnolipids production. From figure 19, it has shown that the initial pH of 7.1 was the best condition for more simultaneous utilization but resulted in a significant increase of pH during cultivation. In this fed-batch process, we faced the problem that the TDA1 F1.1 strain struggled to utilize ethylene glycol. The problem could be the higher terephthalic acid in the medium, the burden of pPS05 plasmid, and its tetracycline marker for rhamnolipids production. Therefore, we continue this fed-batch culture with *an integrated rhlAB* strain called ALE TDA1ge_*rhlAB* + p_*tph*. Since terephthalic acid is a more preferred substrate for this strain than ethylene glycol (see figures 3, 5, 6, and 7), 30 mM of terephthalic was used with the same concentration as ethylene glycol.

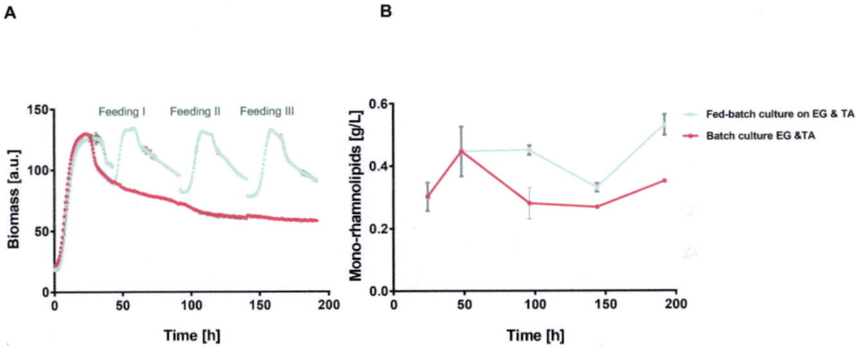

Figure 20 A: Fed-batch culture (blue curve) and control (pink curve) with ethylene glycol 30 mM and terephthalic acid 30 mM as sole carbon sources for **B:** rhamnolipid production. Growth was accessed by growth profiler in 2 ml medium in 24-square-well plates. Errors depict the deviation of the mean (n = 2).

After 48 hours, the first feeding was applied, and in total, three times were conducted during the cultivation. The control without any feedings was also conducted as a comparison. From this fed-batch cultivation in the shake flasks, the rhamnolipid production was achieved from the first fed-batch culture (FB I), second fed-batch culture (FB II), and third fed-batch culture (FB III) with the concentration of 0.45 ± 0.02, 0.33 ± 0.01, 0.53 ± 0.03 g/L, respectively. Overall, fed-batch culture resulted in relatively higher rhamnolipids than batch culture. Surprisingly, biomass also increased by fed-batch culture, while rhamnolipids were stably produced. In the batch culture, rhamnolipids could be consumed as a carbon source when the main substrates are depleted. During cultivation with a feeding system, the pH can also be controlled with the feed; therefore, PET monomer utilization benefits from the sequential fed-batch process. Fed-batch and continuous cultures conducted in fermenters might lead to further improvements on PET monomer utilization.

3.3 PU monomers utilization by evolved *P. capeferrum* TDA1 for rhamnolipid production

Romualdus Nugraha Catur Utomo[1], Till Tiso[1], Herman J. Heipieper[2], Christian Eberlein[2], Lars M. Blank[1]

[1]Institute of Applied Microbiology-iAMB, Aachen Biology and Biotechnology-ABBt, RWTH Aachen University, Aachen, Germany

[2]Department of Environmental Biotechnology, Helmholtz Centre for Environmental Research - UFZ, Leipzig, Germany

Contributions:

1. This chapter was written by Romualdus Nugraha Catur Utomo and reviewed by Lars M. Blank. Romualdus Nugraha Catur Utomo performed cultivation of *P. capeferrum* TDA1 on mixed monomers, toxicity test of ALE *P. capeferrum* TDA1 on mixed monomers containing different concentration of TDA, and rhamnolipid production from PU monomers as substrates.
2. Benjamin Skorupa conducted toxicity and pH tests on adipic acid as a part of his bachelor's thesis (Skorupa, 2020).
3. Till Tiso provided pPS05 and pSK02 for mono-rhamnolipid synthesis.

3.3.1 Toxicity test and pH test on adipic acid

Since ALE TDA1 can grow on all single PU monomers, including adipic acid, 1,4-butanediol, ethylene glycol (figure 13), and 2,4-toluenediamine, it could be an ideal candidate for a consolidated PU upcycling strain. From previous observation, it has been known that *P. capeferrum* TDA1 could utilize well 10 mM of adipic acid as the sole carbon source (Utomo *et al.*, 2020). It would be a great advantage to have adipic acid-utilizing *Pseudomonas* because limited species have been reported so far being able to utilize adipic acid natively (Parke *et al.*, 2001). However, the growth of the wild type and evolved strains (ALE TDA1) on the increased concentrations of adipic acid are unknown and were investigated here. Therefore, a toxicity test was conducted by adding different concentrations of adipic acid when the culture reached its exponential phase. First, 10 mM adipic acid was used as the initial substrate. After that, the cultures were fed with adipic acid to obtain the final concentration of culture of 30 mM, 60 mM, and 90 mM adipic acid, respectively. After approximately 25 hours in the second main culture, both the wild type and the evolved strains could utilize about 30 mM adipic acid in all cultures. No decrease of final biomass was observed when the concentrations of the monomer increased (Skorupa, 2020).

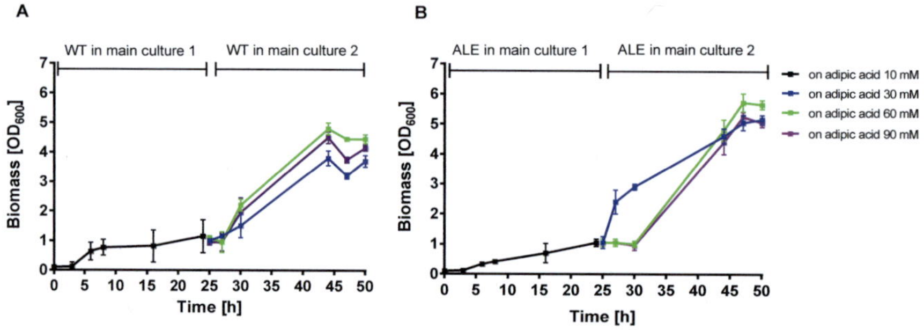

Figure 21 Toxicity test of wild type *P. capeferrum* TDA1 **(A)** and ALE TDA1 **(B)** on adipic acid at different concentration of 30, 60, and 90 mM. Growth was accessed by using the growth profiler in 2 ml medium in 24-square-well plates. Errors depict the deviation of the mean (n = 2) (Skorupa, 2020).

However, both strains can only utilize 30 mM adipic acid within 24-hour incubation. With about 72 mM of buffer in mineral salt medium (MSM), we expected that it could have held the buffer capacity when the culture became acidic as an effect of adipic acid utilization by the strains. Nevertheless, we observed that the culture pH increased when adipic acid was consumed. As an improvement, a pH optimization test was needed to find the best condition for the strain to grow on adipic acid (Skorupa, 2020).

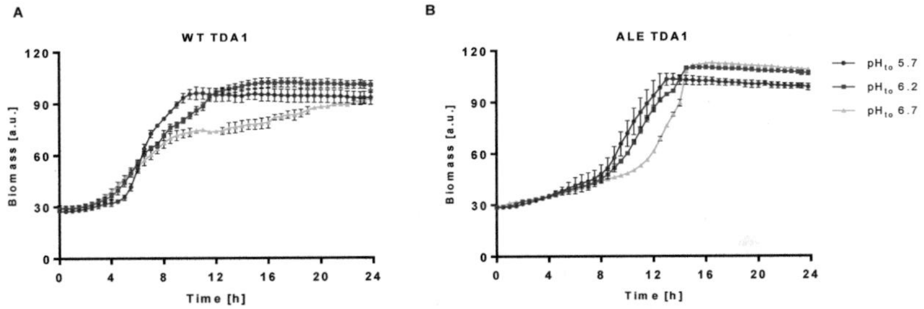

Figure 22 The growth of wild type TDA1 (**A**) and ALE TDA 1 (**B**) on 10 mM adipic acid in MSM with different initial pH of 5.7, 6.2, and 6.7. Growth w60as accessed by growth profiler in 2 ml medium in 24-square-well plates. Errors depict the deviation of the mean (n = 2) (Skorupa, 2020).

The pH is one of the essential factors in soil bacterium growth (Sánchez-Clemente *et al.*, 2018; Cárdenas Espinosa *et al.*, 2020), e.g., if the acidic monomer is used as a carbon source, the pH tends to increase during the culture. As shown in pH optimization, the strain grew well at pH 5.7 – 6.2 to obtain suitable pH during the growth on adipic acid.

3.3.2 Consolidated utilization of PU monomers

The adaptive laboratory evolution of *P. capeferrum* TDA1 in the previous chapter (3.2) enables the strain to grow on ethylene glycol. The strain was grown on a mixture of 10 mM adipic acid, 15 mM 1,4-butanediol, and 30 mM ethylene glycol (with 60 mM carbon equivalent of each). The toxicity test on adipic acid showed that ALE TDA1 is similar to the wild type on adipic acid. Therefore, ALE TDA1 is proposed in the consolidated utilization of PU monomers.

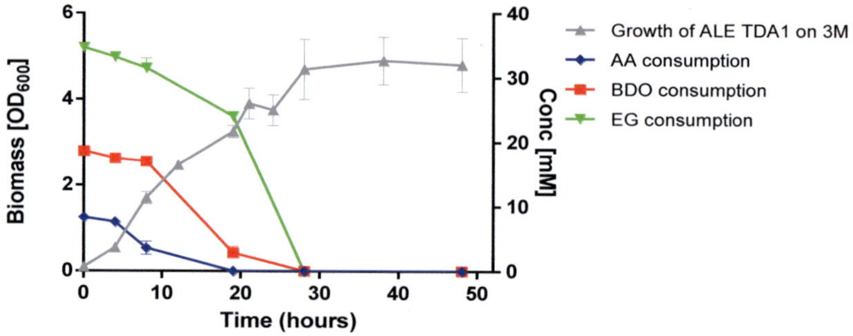

Figure 23 Consolidated utilization of PU monomers by ALE *P. capeferrum* TDA1. Growth was accessed by using the growth profiler in 2 ml medium in 24-square-well plates. Errors depict the deviation of the mean (n = 2).

Based on the analysis of monomer utilization, adipic acid was utilized first with the maximum consumption rate (q_s) of 0.33 ± 0.01 mmol g_{cdw}^{-1} h^{-1} and followed by 1,4-butanediol and ethylene glycol with the consumption rate of 0.62 ± 0.05 and 0.42 ± 0.01 mmol g_{cdw}^{-1} h^{-1}, respectively. Adipic acid is a native substrate of the wild type, which plays a pivotal role in initial growth. Subsequently, 1,4-butanediol was consumed to obtain significant growth at the exponential phase, while ethylene glycol was significantly used after 10 hours.

3.3.3 Toxicity test of evolved *P. capeferrum* TDA1 on 2,4-toluenediamine

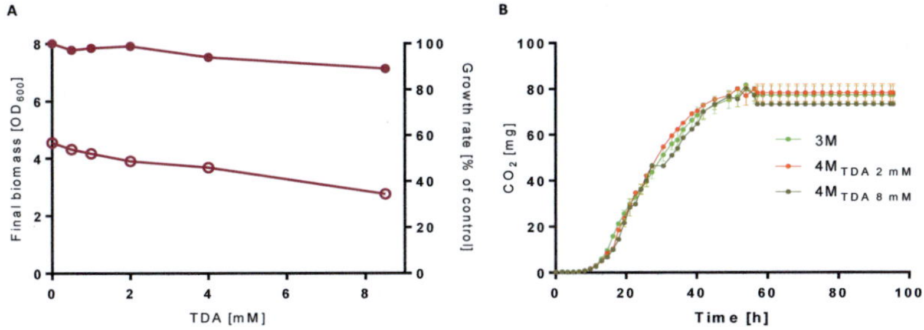

Figure 24 A: Toxicity test of ALE *P. capeferrum* TDA1 on mixed PU monomers with increased 2,4-toluenediamine (2,4-TDA) concentration. Filled circles: effect of 2,4-TDA on the growth rate of ALE *P. capeferrum* TDA1, and empty circles: effect of 2,4-TDA on the biomass of ALE *P. capeferrum* TDA1. **B**: CO_2 emission measured with bluesens. Growth for toxicity test was accessed by growth profiler in 2 ml medium in 24-square-well plates, while for CO_2 off-gas measurement was conducted in 250 ml shake flasks (Duran) connected with gas sensor (Bluesens). Errors depict the deviation of the mean (n = 2).

From toxicity test on increased 2,4-toluenediamine (2,4-TDA) in mixed PU monomers, it has been known that 2,4-TDA concentrations correlated negatively with the rate of growth. Even though 2,4-TDA also shows an inhibitory effect on ALE *P. capeferrum* TDA1, this strain is more tolerant to 2,4-TDA than the defined mixed culture used in the previous study (see chapter 3.1 figure 7). On the other hand, CO_2 off-gas from cultivation on mixed PU monomers with- and without 2,4-TDA showed no difference, suggesting that 2,4-TDA was not converted to CO_2 under these conditions.

3.3.4 Rhamnolipid production from PU monomers

ALE *P. capeferrum* with integrated *rhlAB* – integrated strain, called ALE TDA1 ge_*rhlAB*, was used as a consolidated strain to utilize the mock hydrolysate of PU. ALE TDA1 ge_*rhlAB* was chosen to omit the antibiotic marker and maintain stable and efficient expression of rhamnolipid production during fermentation. After integrating *rhlAB*, the strain lost its ability to utilize adipic acid; therefore, the *dca* operon from *Acinetobacter baylyi* ADP1 was introduced to enable the strain to utilize adipic acid again. The preliminary result of proteomic analysis using adipic acid and glucose feeding in MSM only showed the increased level from

3-hydroxyacyl-CoA dehydrogenase (encoded by *dcaH*), two candidates for 3-oxoadipyl-CoA thiolases (encoded by dcaF), and glycine dehydrogenase with the increased level of enzyme of 3.6, 5.42, 3.15, and 15 folds, respectively (Skorupa, 2020).

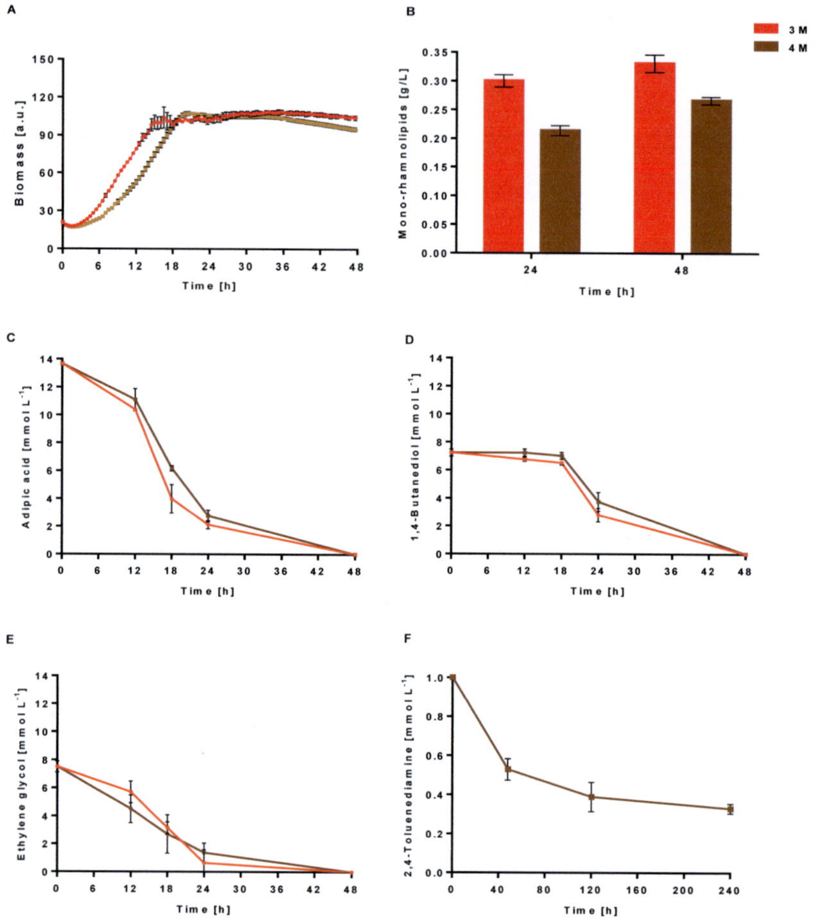

Figure 25 A: Growth of ALE TDA1ge_*rhlAB* on mock hydrolysate of PU monomers with- (4M) and without 2,4-TDA (3M) **B:** the production of rhamnolipids, and the utilization of adipic acid (**C**), 1,4-butanediol (**D**), ethylene glycol (**E**), and 2,4-toluenediamine (2,4-TDA) (**F**). Growth was accessed by growth profiler in 2 ml medium in 24-square-well plates. Errors depict the deviation of the mean (n = 2).

Based on toxicity test on 2,4-TDA, the mock PU hydrolysate consisting of 14 mM adipic acid, 7 mM 1,4-butanediol, 7 mM ethylene glycol, and 1 mM 2,4-toluenediamine was used as sole carbon sources for mono-rhamnolipid production (Songwon, 2017). One millimolar of 2,4-TDA gave a slight inhibitory effect of growth and rhamnolipid production depicted in figure 25. The optimum biodegradation of 2,4-TDA in 4M mixture was around half of 2,4-TDA within 48 hours with the maximum growth rate of 0.18 ± 0.00 h^{-1}, which is lower than the control (3M) (0.21 ± 0.01 h^{-1}). The maximum consumption rate (q_s) of adipic acid, 1,4-butanediol, and ethylene glycol on 4M mixture was slightly lower than on the 3M mixture with the values of 0.55 ± 0.05 to 0.48 ± 0.00, 0.18 ± 0.02 to 0.12 ± 0.03, 0.31 ± 0.03 to 0.26 ± 0.05 mmol g_{cdw}^{-1} h^{-1}, respectively.

From this study, the consolidated strain ALE *P. capeferrum* TDA1 was used as a proof-of-concept in upcycling, which can be applied for the bioconversion of mock hydrolysate of PU to rhamnolipids. Indeed, the maximum biodegradation of 2,4-toluenediamine should be further investigated.

3.4 Upcycling PET and PU monomers to phenazine by engineered *P. capeferrum* TDA1

Romualdus Nugraha Catur Utomo[1], Simone Schmitz[1], Till Tiso[1], Herman J. Heipieper[2], Christian Eberlein[2], Lars M. Blank[1]

[1]Institute of Applied Microbiology-iAMB, Aachen Biology and Biotechnology-ABBt, RWTH Aachen University, Aachen, Germany

[2]Department of Environmental Biotechnology, Helmholtz Centre for Environmental Research - UFZ, Leipzig, Germany

Contributions:

1. This chapter was written by Romualdus Nugraha Catur Utomo and reviewed by Lars M. Blank. Romualdus Nugraha Catur Utomo performed the transformation of phenazine production plasmid to dedicated strain and cultivation for the phenazine production using PET and PU monomers as substrates.
2. Simone Schmitz constructed the plasmid for phenazine production.
3. Jana Fees conducted toxicity test of phenazine.

3.4.1 Phenazine production by *P. capeferrum* TDA1 by using PET monomers as sole carbon sources

After successfully upcycling PET and PU monomers to rhamnolipids, the diversification of valuable products would be the next challenge to increase the value of plastic waste in a circular bioeconomy. To answer this challenge, we proposed producing another valuable product, called phenazine. In this study, ALE TDA1 + p_*tph* + pJNN_*phzAG*_*phzMS* was used to utilize PET monomers and to produce phenazine.

From Chapter 3.2, it was apparent that the utilization of ethylene glycol remained challenging. Therefore, we observed first the best pre-culture condition for the best conversion into biomass using 30 mM ethylene glycol as a sole carbon source in the main culture. Pre-culture consisting of 10 mM terephthalic acid and 10 mM ethylene glycol (equivalent to 2.28 g/L) was chosen as the best pre-culture condition due to its shortest lag phase and maximum growth rate (figure 26 A). On the other hand, only ethylene glycol in the pre-culture resulted in a prolonged lag phase in the main culture.

Since terephthalic acid utilization affects the increased pH during cultivation, we also investigated the optimum initial pH of culture. Different initial pH and substrate could give different results in the growth and production of a particular substance. For example, pH of 5.7 resulted in a more extended lag phase than the other higher pH values in the production of rhamnolipids from PET monomers consisting of ethylene glycol and terephthalic acid (chapter 3.2). In addition, the growth of both wild type and ALE *P. capeferrum* TDA1 on adipic acid was not affected with low initial pH of 5.7 (chapter 3.3). In this study, the growth of ALE TDA1 + p_*tph* + pJNN_*phzAG*_*phzMS* on terephthalic acid with different initial pH was similar. At the same time, the optimum production of phenazine was obtained at an initial pH of 6.7 with a titer of 48.17 ± 0.07 µg/ml (Figures B1 and B2).

Based on the pre-culture test, the mixture of 10 mM ethylene glycol and 10 mM terephthalic acid was chosen as substrates in the pre-culture, and the narrow range of different initial pH (6.4, 6.7, and 7.0) was applied for co-substrate utilization of PET monomers (terephthalic acid and ethylene glycol) and phenazine production.

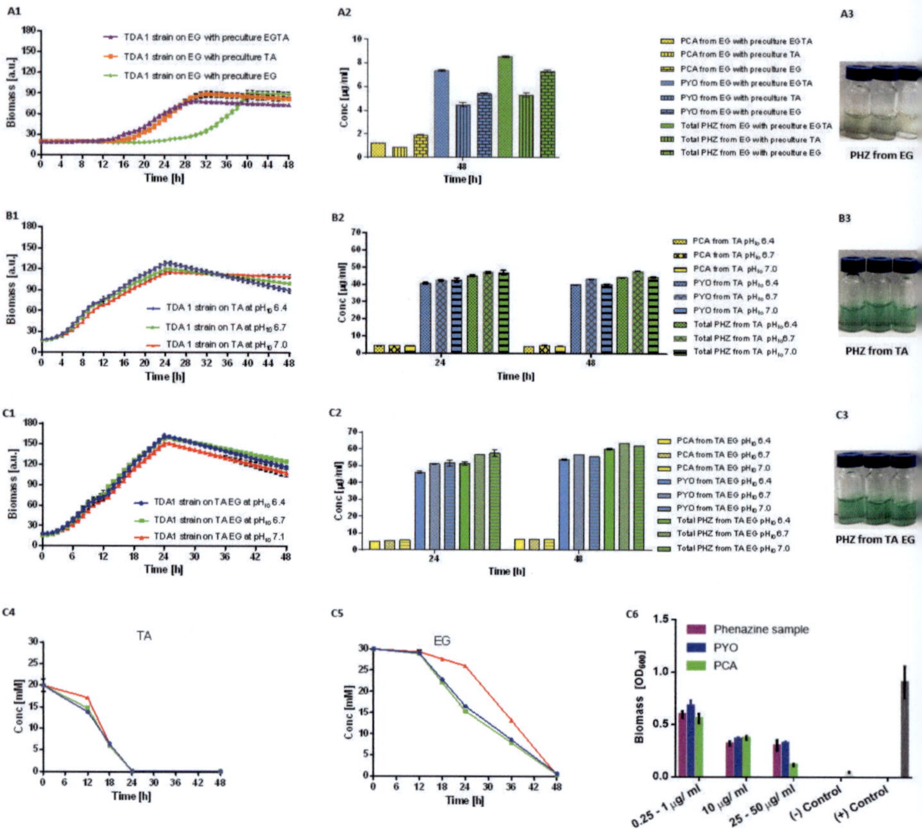

Figure 26 Upcycling of PET monomers for phenazine production **A1:** The growth of engineered *P. capeferrum* TDA1 on 30 mM ethylene glycol with different pre-culture conditions. **A2:** Phenazine production by engineered *P. capeferrum* TDA1 with 30 mM ethylene glycol in MSM. **A3:** Phenazine production from ethylene glycol as a sole carbon source. **B1:** The growth of engineered *P. capeferrum* TDA1 on 20 mM terephthalic acid with different initial pH. **B2:** Phenazine production by engineered *P. capeferrum* TDA1 with 20 mM terephthalic acid in MSM adjusted to different initial pH. **B3:** The visual of phenazine production from terephthalic acid as a sole carbon source. **C1:** The growth of engineered *P. capeferrum* TDA1 on 20 mM terephthalic acid and 30 mM ethylene glycol with different initial pH. **C2:** Phenazine production by engineered *P. capeferrum* TDA1 with 20 mM terephthalic acid and 30 mM ethylene glycol in MSM adjusted to different initial pH. **C3:** The visual of phenazine production

from terephthalic acid and ethylene glycol as sole carbon sources. **C4 and 5:** The uptake of terephthalic acid and ethylene glycol during culture time. **C6** Toxicity test of different concentrations of phenazine sample from engineered *P. capeferrum* TDA1 (purple), commercial-grade of pyocyanin (PYO/ blue), and phenazine 1-carboxylic acid (PCA/ green) on *Saccharomyces cerevisiae* in yeast extract peptone dextrose (YPD) medium. Growth for phenazine production was accessed by growth profiler in 2 ml medium in 24-square-well plates, while plate reader accessed growth from toxicity test of phenazine in 50 µL medium in F-bottom 96-well plates. Errors depict the deviation of the mean (n = 2).

From the result, the growth trends on the initial pH of 6.4 and 6.7 were relatively similar with the maximum growth rate (µ max) of 0.37 ± 0.01 h^{-1} and slightly higher than the initial pH of 7.0 with the growth rate of 0.36 ± 0.01 h^{-1} (figure C1). It is also shown that the optimum production of phenazine from mixed PET monomers was obtained at pH of 6.7 with a concentration of 63.46 ± 0.06 µg/ml after 48-hour incubation (figure C2). In addition to this, the initial pH of 6.7 also led to the most simultaneous utilization of terephthalic acid and ethylene glycol with the maximum consumption rate (q_s) of 0.51 ± 0.01 and 0.37 ± 0.00 mmol g_{cdw}^{-1} h^{-1}, respectively (figure C4 and 5). According to ISO 20776-1:2019, a toxicity test of phenazine was conducted on *Saccharomyces cerevisiae* to see its inhibitory effect (ISO 20776-1:2019(E), 2019). From the toxicity test, it is shown that the concentration of 0.25-1 µg/ml of phenazine sample and commercial grade of PYO and PCA gave slight inhibition towards the growth of *S. cerevisiae.* Afterwards, the biomass was around 60% reduced in the presence of phenazine at a concentration of 10 µg/ml. The concentration between 25-50 µg/ml of phenazine samples and PYO commercial grade did not give a further significant reduction of biomass. However, PCA commercial grade could inhibit the growth of *S. cerevisiae* by 87% reduction of initial biomass.

3.4.2 Phenazine production by using PU monomers by *P. capeferrum* TDA1

Previous chapters have established the conversion of mock hydrolysate of PU to rhamnolipids by mixed and single cultures (chapters 3.1 and 3.3). In this study, diversification of high-value products from mock hydrolysate of PU was conducted by enabling ALE TDA1 strain to produce phenazine via transformation of pJNN_*phzAG_phzMS*. A similar approach in chapter 3.3, 14 mM adipic acid, 7 mM 1,4-butanediol, 7 mM ethylene glycol, and 1 mM 2,4-TDA was chosen as sole carbon sources.

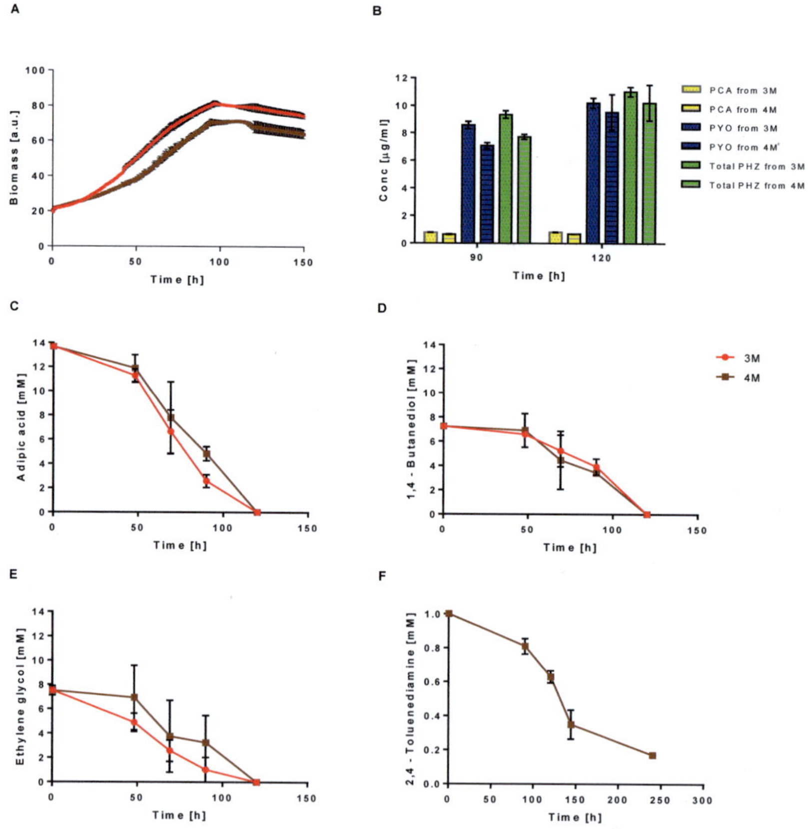

Figure 27 A: Growth of ALE *P. capeferrum* TDA1 pJNN phzA1-G + phz MS on mock hydrolysate of PU monomers with- (4M) and without 2,4-TDA (3M) corresponds to **B:** the production of phenazine, and the utilization of adipic acid (**C**), 1,4-butanediol (**D**), ethylene glycol (**E**), and

2,4-toluenediamine (2,4-TDA) (**F**). Growth was accessed by growth profiler in 2 ml medium in 24-square-well plates. Errors depict the deviation of the mean (n = 2).

The optimum biodegradation of 2,4-TDA in 4M mixture was only around 40% within 120 hours, while the other monomers, namely adipic acid, 1,4-butanediol, and ethylene glycol, had been completely utilized. 4M mixture had a lower maximum growth rate of 0.05 ± 0.00 h^{-1} than the control mixture without the addition of 2,4-TDA (3M), which enabled a growth rate of 0.06 ± 0.00 h^{-1}. Even though 2,4-TDA had not been completely degraded within 120 hours, the maximum consumption rates (q_s) of adipic acid in 4M and 3M mixture were comparable at 0.19 ± 0.01 mmol g_{cdw}^{-1} h^{-1}, even q_s of 1,4-butanediol was slightly higher than in 3M mixture with the values of 0.08 ± 0.01 to 0.06 ± 0.00 mmol g_{cdw}^{-1} h^{-1}. Only the maximum consumption rate (q_s) of ethylene glycol in the 4M mixture was slightly lower than in the 3M mixture with the value of 0.09 ± 0.01 to 0.11 ± 0.01 mmol g_{cdw}^{-1} h^{-1}.

3.5 Exploit the bow-tie structure of engineered *P. capeferrum* TDA1 for rhamnolipid and phenazine production from plastic monomers

Romualdus Nugraha Catur Utomo[1], Simone Schmitz[1], Till Tiso[1], Herman J. Heipieper[2], Christian Eberlein[2], Lars M. Blank[1]

[1]Institute of Applied Microbiology-iAMB, Aachen Biology and Biotechnology-ABBt, RWTH Aachen University, Aachen, Germany

[2]Department of Environmental Biotechnology, Helmholtz Centre for Environmental Research - UFZ, Leipzig, Germany

Contributions:

1. This chapter was written by Romualdus Nugraha Catur Utomo and reviewed by Lars M. Blank. Romualdus Nugraha Catur Utomo designed and performed the bow-tie structure experiment.
2. Simone Schmitz constructed the plasmid for phenazine production.
3. Till Tiso provided pPS05 and pSK02 for mono-rhamnolipid synthesis.

After successful upcycling of PET and PU monomers, the consolidated strain P. *capeferrum* TDA1 was used for an experiment mimicking household waste or the mixture of waste in the environment, including monomers representing aromatic polymers (e.g., lignin, plastics), cellulose, C-C bond plastics, PET, and PU. The model monomers chosen were benzoate, glucose, decanoate, adipic acid, 1,4-butanediol, 2,3-butanediol, ethylene glycol, and terephthalic acid. Such a mixture might mimic waste hydrolysates and the situation in contaminated niches in the environment, where microorganisms might play essential roles in the biodegradation of various wastes. However, the challenge is converting the polymers into monomers under those conditions. The scenario to clean the environment is discussed. However, the technical challenges and the public concern to finally overcome the word "containment" in synthetic biology by releasing engineered strains would be a massive change in technology security and argumentation. Finally, the evolved and engineered *P. capeferrum* TDA1 have been applied for rhamnolipid (ALE TDA1ge_*rhlAB*_evAA & ALE TDA1ge_*rhlAB* + p_ *tph*) and phenazine (ALE TDA1evAA + pJNN_*phzAG*_*phzMS* & ALE TDA1 + p_*tph* + pJNN_*phzAG*_*phzMS*) productions under these unconventional conditions.

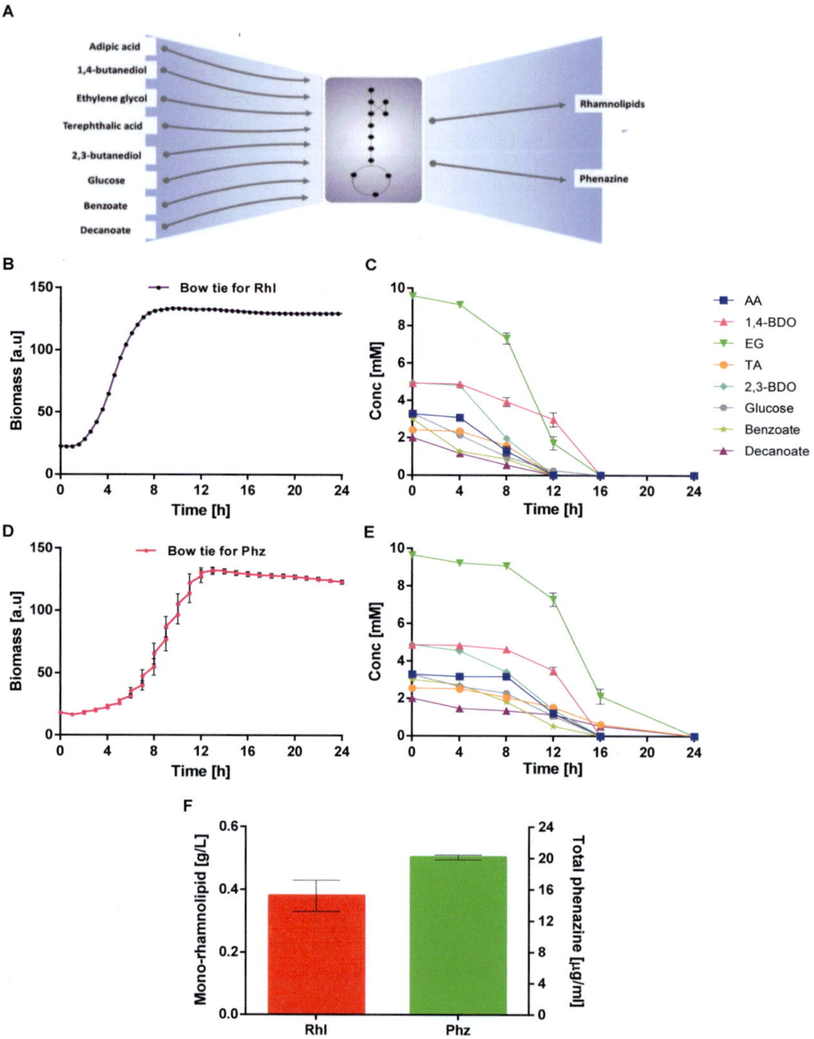

Figure 28 A: Bow-tie structure from various substrates to value-added products. Producing rhamnolipids and phenazine from 8 different substrates, namely adipic acid, 1,4 butanediol, ethylene glycol, terephthalic acid, 2,3-butanediol, glucose, benzoate, and decanoate with 20 mM carbon equivalent (160 mM carbon in total). Growth of ALE and engineered TDA1 strains for **B**: rhamnolipids and **D:** phenazine production using **C and E**: mixed substrates of bow-tie

structure as sole carbon sources, respectively **F:** Final concentration of mono-rhamnolipid and total phenazine from bow-tie structure exploitation. Growth was accessed by growth profiler in 2 ml medium in 24-square-well plates. Errors depict the deviation of the mean (n = 2).

From exploiting bow-tie structure to high-value products, namely mono-rhamnolipids and phenazine, the maximum bacterial growth rates were 0.65 ± 0.02 and 0.22 ± 0.02 h^{-1}, respectively. With those growth rates, all monomers, except 1,4-BDO and EG, were completely consumed within 12 hours for mono-rhamnolipid production. However, it needed at least 16 hours to consume 1,4-BDO and EG for phenazine production. This study successfully obtained mono-rhamnolipids with a final titer of 0.38 ± 0.05 g/L and total phenazine with a final titer of 20.13 ± 0.3 µg/ml.

As an evaluation, pJNN *phzA-G phz MS* plasmid construct with a large size of 14.2 kbp could be a burden for bacterial cells and cause a lower growth rate in phenazine production, while in mono-rhamnolipids production, the two responsible genes (the A and B) had been integrated into the genome. The integrated *phzA-G phz MS* into the genome might give a better growth rate and production of phenazine.

Table 16 Consumption rate of each monomer from exploitation of bow-tie structure for rhamnolipid and phenazine productions

Polymer	**Monomer**	**q_s (mmol g_{cdw}^{-1} h^{-1})**	
		for Rhl production	For phenazine production
PU	Adipic acid	0.63 ± 0.01	0.48 ± 0.00
PU	1,4-butanediol	0.38 ± 0.06	0.31 ± 0.04
PU/ PET	Ethylene glycol	1.51 ± 0.06	0.54 ± 0.08
PET	Terephthalic acid	0.46 ± 0.00	0.23 ± 0.00
PU	2,3-butanediol	0.95 ± 0.02	0.79 ± 0.00
Cellulose	Glucose	0.58 ± 0.00	0.51 ± 0.03
Lignin	Benzoate	0.58 ± 0.00	0.56 ± 0.00
Lipid	Decanoate	0.38 ± 0.00	0.20 ± 0.01

3.6 *P. capeferrum* TDA1 as plastic monomer utilizer and bioplastic producer

Romualdus Nugraha Catur Utomo[1], Herman J. Heipieper[2], Christian Eberlein[2], Lars M. Blank[1]

[1]Institute of Applied Microbiology-iAMB, Aachen Biology and Biotechnology-ABBt, RWTH Aachen University, Aachen, Germany

[2]Department of Environmental Biotechnology, Helmholtz Centre for Environmental Research - UFZ, Leipzig, Germany

Presented and awarded as the best conference paper at The International Conference on Environmental and Sustainability Context (ICoESCo)
and published as: "New species *Pseudomonas capeferrum* TDA1 as a plastic monomer utilizer and a PHA native producer." *IOP Conf. Series: Earth and Environmental Science* 2021, 894, 012024. DOI: 10.1088/1755-1315/894/1/012024.

Contributions:

1. This chapter was written by Romualdus Nugraha Catur Utomo and reviewed by Lars M. Blank. Romualdus Nugraha Catur Utomo performed cultivations, and qualitative analysis of PHA granule formation.
2. This chapter was partly reviewed by Christian Eberlein and Hermann J. Heipieper.

3.6.1 Utilization of plastic monomer and other preferable substrates by *P. capeferrum* TDA1

Previously, *P. capeferrum* TDA1, which can utilize adipic acid (AA), 1,4-butanediol (1,4-BDO), and 2,4-toluenediamine (2,4-TDA), was evolved to be able to utilize ethylene glycol (Cárdenas Espinosa *et al.*, 2020; Utomo *et al.*, 2020). At the end of 19-day adaptive laboratory evolution, this strain can utilize 30 mM of ethylene glycol (EG) with OD_{600} of 1.8 (see chapter 3.2) and is now called ALE TDA1 strain. It has been proved in many studies that PHA is produced by bacteria when the environmental condition is not optimal due to the limitation of nitrogen or other nutrients and an excess of carbon source. Furthermore, nitrogen limitation plays a vital role in enhanced PHA granule formation in most Pseudomonads. Therefore, we first observed the growth of the ALE TDA1 strain on several PU monomers (AA, 1,4-BDO, and EG) and medium-chain fatty acids (decanoate and octanoate) as a sole carbon source. At the same time, one-tenth of the usual amount of ammonium sulfate solution was used as a limited nitrogen source (Ciesielski *et al.*, 2010; Poblete-Castro, Escapa, *et al.*, 2012). Decanoate and octanoate were used in this study as two of the best alternative substrates in PHA production.(Poblete-Castro, Escapa, *et al.*, 2012; Borrero *et al.*, 2014) Each substrate used in this experiment, gave equal C-molar concentration of 0.333 mol/L with 55 mM AA, 83 mM 1,4-BDO, 166 mM EG, 33 mM decanoate, and 42 mM octanoate.

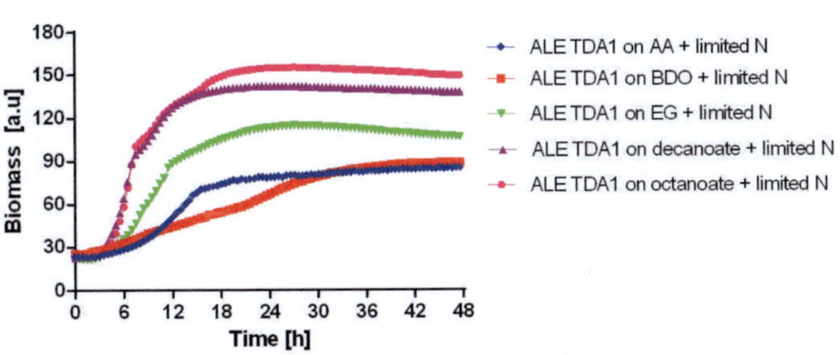

μ max (h^{-1})
0.12 ± 0.01
0.06 ± 0.01
0.20 ± 0.00
0.34 ± 0.01
0.38 ± 0.01

Figure 29 The growth of ALE TDA1 in MSM with different single monomers as a sole carbon source. Growth was accessed by growth profiler in 2 ml medium in 24-square-well plates. Errors depict the deviation of the mean (n = 2) (Utomo *et al.*, 2021).

Based on the growth on AA, 1,4-BDO, EG, decanoate, and octanoate with nitrogen limitation, we could see that growth on EG gave the highest maximum growth rate and the highest biomass among plastic monomers. On the other hand, growth on decanoate and octanoate outperformed growth on plastic monomers with maximum growth rates of 0.34 and 0.38 h^{-1}, respectively. As preferable and unique alternative substrates, decanoate, octanoate, and a mixture of fatty acids can be utilized by Pseudomonads, for instance, *P. putida* KT2440 and *P. putida* mt-2 for mcl-PHAs production (Poblete-Castro, Escapa, *et al.*, 2012; Borrero-de Acuña *et al.*, 2019).

3.6.2 PHA granule formation from plastic monomer and other preferable substrates as sole carbon sources

Even though plastic monomers were less preferred as sole carbon sources than the reference substrates (decanoate and octanoate), we further investigated the growth of *P. capeferrum* TDA1 on the mixture of PU monomers with 10 mM AA, 15 mM 1,4-BDO, and 30 mM EG (60 C-mM of each). The mixture of PU monomers with- and without nitrogen limitation was used as substrates to see the growth behavior of *P. capeferrum* TDA1 and consumption rate on those two different conditions. Since more 2,4-TDA gave a toxicity effect, only 1 mM of 2,4-TDA was used in this study to see the minimum effect on both 4M and 4M + nitrogen limitation medium.

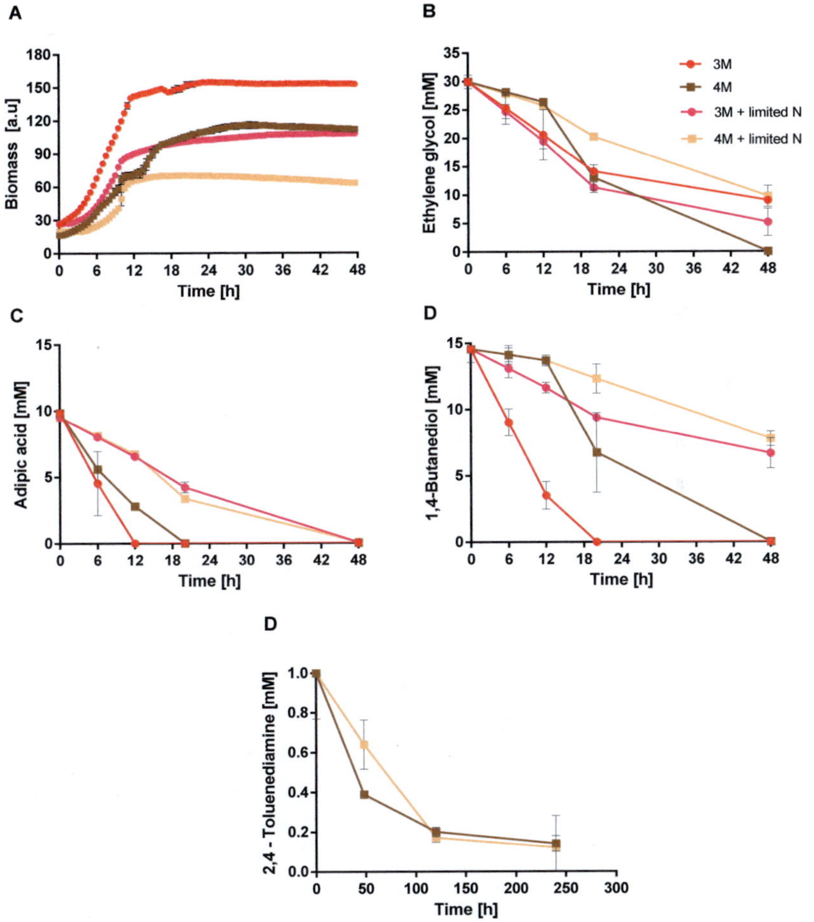

Figure 30 The growth of ALE TDA1 in MSM with 3M (AA, 1,4-BDO, and EG), 4M (AA, 1,4-BDO, EG, 2,4-TDA), 3M + nitrogen limitation, and 4M + nitrogen limitation as sole carbon sources. Growth was accessed by growth profiler in 2 ml medium in 24-square-well plates. Errors depict the deviation of the mean (n = 2).(Utomo *et al.*, 2021)

It is depicted in figure 30 that nitrogen limitation in 3M and 4M lowered the biomass than the control even though the maximum growth rate between with- and without nitrogen limitation was similar (table 17). The lower biomass showed that nitrogen source plays a vital role in the growth of the ALE TDA1 strain. Nitrogen source is essential for the growth and biochemical synthesis of nucleic acids, lipids, and proteins in bacteria. Nitrogen also regulates product formation, in which the biosynthesis of secondary metabolites is limited by the nitrogen sources favoring cell growth, such as ammonium sulfates and certain types of amino acids. Biodegradation of 2,4-TDA was only observed until 120 h incubation by ALE TDA1, showing better utilization in 4M than in 4M with nitrogen limitation.

Table 17 Effect of nitrogen limitation on MSM medium with 3M and 4M as sole carbon sources towards growth of *P. capeferrum* TDA1 and consumption rates of each monomer(Utomo *et al.*, 2021)

Substrate	**µ Max (h^{-1})**	**q_s (mmol g_{cdw}^{-1} h^{-1})**	
		AA	: 1.48 ± 0.24
3M	0.20 ± 0.00	1,4-BDO	: 2.13 ± 0.04
		EG	: 1.81 ± 0.02
		AA	: 1.29 ± 0.06
4M	0.23 ± 0.00	1,4-BDO	: 1.02 ± 0.24
		EG	: 2.22 ± 0.08
		AA	: 1.02 ± 0.01
3M with nitrogen limitation	0.21 ± 0.00	1,4-BDO	: 0.99 ± 0.15
		EG	: 1.68 ± 0.11
		AA	: 1.51 ± 0.09
4M with nitrogen limitation	0.21 ± 0.01	1,4-BDO	: 0.55 ± 0.04
		EG	: 1.19 ± 0.06

It is also attractive to see AA, 1,4-BDO, and EG consumption rates in those various mixtures. While the consumption of AA and EG was relatively similar in 3M and 4M, the consumption rate of 1,4-BDO was lower in 4M and further lower in 3M with nitrogen limitation and 4M with nitrogen limitation, respectively. Even though few studies of bioconversion of synthetic plastic into bioplastic, the prominent studies are from upcycling PET to PHA. From Kenny et al., initially, PET was pyrolyzed at 450°C resulting solid fraction of terephthalic acid, which *P. umsongensis* GO16 used as a substrate for PHA production (Kenny *et al.*, 2008, 2012). In a further study, it was genetically identified that responsible genes involved in ethylene glycol metabolism also exist in *P. umsongensis* GO16. After the strain was validated to use terephthalic acid and ethylene glycol as sole carbon sources, the PHA production from both monomers was also proved (Narancic *et al.*, 2021).

As proof of concept, we continued to investigate the PHA granule formation in nitrogen-limited medium with plastic monomers and medium-chain fatty acids as sole carbon source(s) via qualitative analysis, namely plate fluorescence and microscopy analysis.

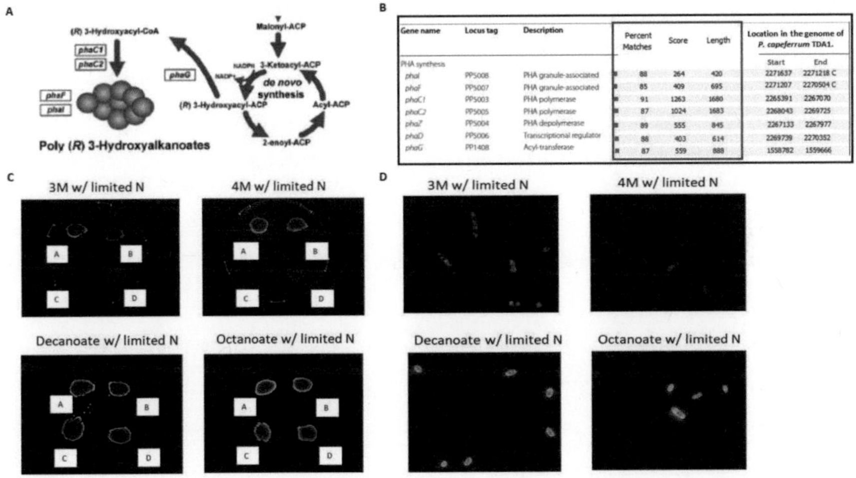

Figure 31 Proof of concept of *P. capeferrum* TDA1 as a native PHA producer. **A:** Biosynthetic pathway of PHA production in *P. putida* KT2440 as a reference.(Beckers *et al.*, 2016) **B:** BLAST results of PHA encoding genes between *P. putida* KT2440 and *P. capeferrum* TDA1. **C:** Plate fluorescence staining for PHA production from 3M (adipic acid, 1,4-butanediol, and ethylene glycol), 3M+2,4-TDA (4M), alternative substrate namely decanoate and octanoate as sole carbon source with nitrogen limitation with different pre-cultures, namely A: glucose, B: glucose + limited N, C: decanoate + limited N, D: 3M + limited N. **D:** Microscopy observation of PHA granule formation from different substrate (Utomo *et al.*, 2021).

Qualitative measurement for PHA production was conducted with a viable - colony staining method using Nile red for direct screening of PHA accumulation (Spiekermann *et al.*, 1999). From figure 31 C, pre-culture in MSM with different substrates were used, namely glucose without nitrogen limitation (A), glucose with nitrogen limitation (B), decanoate with nitrogen limitation (C), and 3M with nitrogen limitation (D). After that, the growing cultures were streaked out on Nile red agar plate with 3M (10 mM adipic acid, 15 mM 1,4-butanediol, and 30 mM ethylene glycol), 4M (3M + 1 mM 2,4-TDA), 55 mM glucose, 33 mM decanoate, and 42 mM octanoate as sole carbon sources. The growth and fluorescent emission were then observed after 1-day incubation.

The PHA production on 3M and 4M mediums was observed when glucose was used as a sole carbon source in the pre-culture, while PHA production on decanoate and octanoate was shown from all pre-culture conditions. It can be concluded that the PHA production from PU monomers needs more favorable conditions in pre-culture. The PHA granule-forming cells were found from all conditions, even after adding 1 mM 2,4-TDA. However, the granules inside the cell were relatively less than the other conditions. 2,4-TDA might inhibit not only the growth but also relatively too little carbon to be taken up (Spiekermann *et al.*, 1999).

According to the study from Beckers et al. (2016), several genes are responsible for PHA granule formation in *P. putida* KT2440, namely *phaC1* and *C2*, for encoding pha polymerases, *phaZ* for encoding pha depolymerase, *phaD* for encoding transcriptional regulator, *phaG* for encoding acyltransferase, and *phaI* and *F* for encoding PHA-forming enzymes. By aligning the local similarity between two molecules (*P. putida* KT2440 and *P. capeferrum* TDA1 genomes) using clone manager (Sci-Ed software), it is confirmed that the responsible genes for PHA synthesis are also present in *P. capeferrum* TDA1.

3.7 Prospective future of *P. capeferrum* TDA1 in industrial and environmental applications

Romualdus Nugraha Catur Utomo[1], Herman J. Heipieper[2], Christian Eberlein[2], Lars M. Blank[1]

[1]Institute of Applied Microbiology-iAMB, Aachen Biology and Biotechnology-ABBt, RWTH Aachen University, Aachen, Germany

[2]Department of Environmental Biotechnology, Helmholtz Centre for Environmental Research - UFZ, Leipzig, Germany

Contributions:

1. This chapter was written by Romualdus Nugraha Catur Utomo and reviewed by Lars M. Blank.
2. Romualdus Nugraha Catur Utomo performed the cultivation of *P. capeferrum* TDA1 at elevated temperatures.

In many cases, biotechnology companies struggle for the cost efficiency of energy consumption, mainly electricity and steam generation. Both are usually needed in fermentation with giant fermenters for producing specific products. Beyond being a biotechnological agent, a robust microorganism would be the answer to that problem. To meet the requirement, the microorganism should be adaptable in a relatively higher temperature than the typical temperature of 30°C. For example, *Escherichia coli* is widely used in industry and a promising host for producing biobased adipic acid, organic acids, and alcohols. Those applications are due to their feasibility for metabolic engineering, and the bacteria can grow well at 37°C (Yu *et al.*, 2014; Liu *et al.*, 2015; Skoog *et al.*, 2018).

Therefore, the TDA1 strain was also tested at a temperature of 30°C, 37°C, 40°C, and 50°C to see its robustness. As a result, the strain grew well at 30°C and 37°C, and this strain was still able to grow at 40°C with lower biomass. Nevertheless, a temperature of 50°C was detrimental for the TDA1 strain.

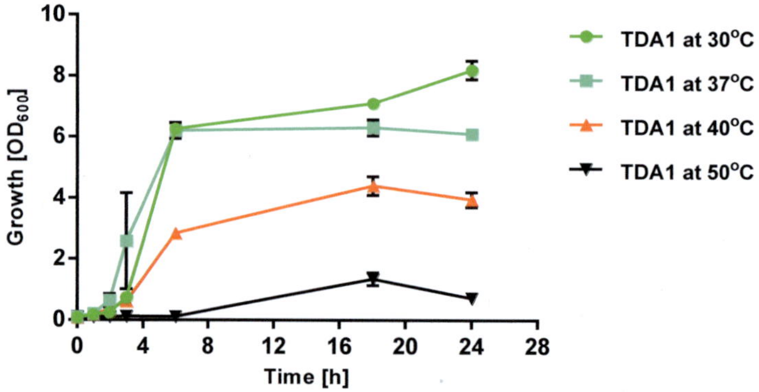

Figure 32 Growth of *P. capeferrum* TDA1 at elevated temperatures. Growth was conducted with 250 ml shake flask. Errors depict the deviation of the mean (n = 2).

Chapter 4

Discussion

Contributions:

This chapter was written by Romualdus Nugraha Catur Utomo and revised by Lars M. Blank.

4.1 Mixed microbial cultures for mixed substrate utilization

The here presented results of a defined microbial mixed culture utilizing PU monomers contribute to the general discussion of possible strategies to utilize mixed carbon sources (chapter 3.1). In waste-water treatment and biogas production, undefined microbial consortia fulfill the catalytic tasks, importantly, robustly over months and years with ever-changing substrates and sometimes changing physical parameters such as temperature. In such conditions, a complex food web exists, where many (all) of the ecological niches are occupied by consortia members.

Defined mixed cultures can either be organized as presented here, where the catalytic tasks are distributed to the individual members, i.e., the conversion of the different monomers are fulfilled by dedicated consortia members – with a single objective, here, to synthesize rhamnolipids. Such a mixed culture can also be designed that the individual members depend on each other. For example, in the study of Schlembach et al. (2020), cellulose hydrolysis and product synthesis (itaconic acid as an example) were distributed to *Trichoderma reesei* and an engineered *Ustilago maydis* strain, respectively.

Finally, substrate conversion and product formation tasks can be consolidated into a single strain. While for decades such strains were aimed in lignocellulosic biotechnology to, e.g., for bioethanol production, no such strain until now was competitive (Liu *et al.*, 2020). Hence, we separate polymer degradation and monomer consumption (Wierckx *et al.*, 2015). The advantage of a single strain for mixed substrate utilization is the flexibility of central carbon metabolism under ever-changing substrate ratio conditions, i.e., as encountered in syngas fermentation at steel mill sites (Molitor *et al.*, 2016). In this study, the consolidation of three or even four substrate degradation pathways was challenging on its own (see chapters 3.1 and 3.3); engineering true co-utilization at industrially required rates is the next challenge to overcome. That this is not a trivial task can be seen by the 30 years of research on glucose/xylose co-utilization in baker's yeast (Li, X., Chen, Y., & Nielsen, 2019).

4.2 Diversification of high-value products from plastic waste upcycling

Since upcycling is a much-needed alternative to common recycling by adding more value to the carbon-rich plastic waste, diversification of potentially reachable products should be considered. The more value-added products can be synthesized, the more market, industry, and society can engage. Some beneficial and valuable products should be revealed to boost the progress of plastic waste upcycling. Rhamnolipids, intensively discussed in the previous chapters, are prominent products of plastic waste upcycling. After strain construction, the biggest challenge of rhamnolipid production during scale-up is the massive foam formation. A range of technologies has tackled this problem, including the use of antifoam (Beuker *et al.*, 2016), bubble-free membrane aerator (Bongartz *et al.*, 2021), ethanol as carbon and antifoam agent (Bator *et al.*, 2020), second phase fermentations (Blesken, Bator, *et al.*, 2020), and uncoupled foam fractionation (Blesken, Strümpfler, *et al.*, 2020)

Polyhydroxyalkanoates (PHA) are especially attractive for plastic waste upcycling because the bioplastic feeds immediately back into the plastic economy. However, with a clear biodegradability option now, including the advantage of biodegradable plastic blends, a plastic to bioplastic process can be envisioned (Narancic *et al.*, 2018). Many Pseudomonads can natively produce PHA as their intracellular product (Poblete-Castro, Becker, *et al.*, 2012), and it is also shown in chapter 3.6 that strain *P. capeferrum* TDA1 can natively produce PHA. For sure, many improvements are still needed to bring the "plastic to bio-plastic" process on an industrial scale, with the product's price being an ultimate determinant. A life-cycle assessment (LCA) on a PE-based feed for PHA synthesis points out the many challenges one faces when implementing this strategy (Guzik *et al.*, 2021).

Previous studies showed that phenazine is a natural redox mediator in bio-electrochemical systems, potentially in waste-water treatment (Schmitz *et al.*, 2015). This compound was also used in a bio-electrochemical system to reduce the oxygen demand during rhamnolipid production by enabling the whole-cell catalyst to use the anode as a sink for their metabolically derived electrons (Askitosari *et al.*, 2019, 2020). In this study, the plastic monomer-derived phenazine was used in an antimicrobial test (chapter 3.4) to add another application to the repertoire of this compound. Phenazine had some antimicrobial activity at relatively low concentrations; therefore, this substance is discussed as an antimicrobial agent for the food industry as an example(Hamad *et al.*, 2020). Inspired by the microbial warfare strategy, the investigation of bromophenazine showed its efficacy against *Staphylococcus aureus* and *Staphylococcus epidermis*, which are notorious for their drug resistance in humans (Borrero *et al.*, 2014). Phenazine was also effective against mycoplasma, which can cause infections in animal husbandry (Valentine-King *et al.*, 2020). The efficacy of phenazine derivatives as antimicrobial agents against Gram-positive (methicillin-resistant *Staphylococcus aureus*, MRSA) and Gram-negative (*Escherichia coli*) bacteria were reported (Krishnaiah *et al.*, 2018). The compound 6,9-dichloro-*N*-(methylsulfonyl)phenazine-1-carboxamide was the most active molecule against MRSA in this study, with inhibition of biofilm formation of *S. aureus*, whereas 9-methyl-*N*-(methylsulfonyl)phenazine-1-carboxamide showed good activity against both MRSA and *E. coli* (Krishnaiah *et al.*, 2018)

4.3 The plastic waste management in the EU and ASEAN

The mismanagement of plastic waste is high in low- and middle-income countries, such as the top ten global plastic waste contributors in the ASEAN region. There is no significant management towards hazardous and recalcitrant plastic waste from PU, PP, PS, PVC, and polycarbonate, which can endanger the ocean ecosystem. For example, open dumping and burning solid waste, including plastic, is still prevalent in ASEAN countries (Modak *et al.*, 2017). Therefore, better plastic waste management should be encouraged by targeted regulations, comprehensive collaboration among stakeholders, and efficient jurisdiction (McIlgorm *et al.*, 2009; Jain, 2017).

On the other hand, the European Commission shapes European laws and their application to plastic contamination (European Commission, 2019). Based on that act, the regulation is mainly purposed to conserve natural resources and protect human health and the environment during the generation and management of waste. The act is further differentiated by the waste management act of regional states. All regional states have the jurisdiction to add laws that German Federal Laws do not already regulate. Furthermore, the states' laws tend to focus on the implementation of determining subjects for waste disposal obligation, authorized bodies, and municipal waste disposal ordinances. At the municipal level, the ordinances concern more in garbage collection, usage, and integration into the public system. The separations between organic waste, paper, metal, plastic, and glass are mandatory to promote recycling and increase the recovery rate (Umweltbundesamt, 2014).

On the other side, there is a lack of information regarding the agreement of waste regulatory framework and long-term strategy to reduce plastic waste of ASEAN members. In contrast, each ASEAN member has varied strategies and regulations of waste management (Jain, 2017). It could be concluded that most ASEAN members have the "Basic Act on Environment," which regulates air, water, environment, and waste management. However, based on policy, program, and plan assessments, there are prominent drawbacks, such as all regulations are not integrated and implemented well. On the other hand, regarding technology assessment, except in Singapore, the other members lacked waste separation and recycling technology. In contrast, a lack of data and information on waste management and its accessibility becomes a prominent negative aspect of institutional assessment. Urban local bodies (ULBs) and the private sector (formal) should be reinforced to be more involved in programs, plans, and projects for better implementation of the regulation and technology (Jain, 2017).

Mainly, Indonesia showed poor cooperation among stakeholders to tackle the plastic waste problem and was very poor in implementing at all government levels, resulting in waste segregation and low recycling. Based on UN recommendation of waste management in the ASEAN region, there are three essential points: Policy and Regulatory; Institutional, Technical, and Performance; and Funding, Financing, and Economics. Since Indonesia is the most plastic waste contributor in ASEAN, the government must consider improving those recommendations (Jain, 2017; Modak *et al.*, 2017).

Separation of different waste starting from the household is crucial to reduce plastic waste contamination significantly. In Germany, separation is conducted based on different waste substances with different bag colors. For example, the yellow bag or bin is used for plastic waste; the blue one collects paper waste, while the green is used for organic waste such as food waste. That kind of approach should be established in a society in which ULBs, and private sectors are mainly responsible for the collection schedule, distribution of waste bags, and utilization of waste based on the types of waste. Pure organic waste could be easily composted, while paper waste is efficiently recycled. Furthermore, plastic waste could be significantly recycled by physical, chemical, biological approaches, or a combination of them (BMU, 2018), while the residual material mixture can be incinerated for active recovery.

Chapter 5

Conclusion and outlook

Contributions:

This chapter was written by Romualdus Nugraha Catur Utomo and revised by Lars M. Blank.

Many abundant and cheap carbon sources for microbes are not single molecules, such as glucose, but rather a mixture of molecules, such as molasses or lignocellulose hydrolysates, containing some molecules that are either difficult to degrade or just toxic outright. The same is true for PU hydrolysates, containing a mixture of diols, dicarboxylates, and diamines. The microbial utilization of such a hydrolysate was investigated in chapter 3.1, in which a mixture of engineered substrate-specialized *P. putida* strains was used. While 1,4-BDO, EG, and AA are readily used for growth, the isocyanate derivative 2,4-TDA inhibited the growth of the mixed culture, even when a strain that can degrade this diamine was added. The reactive extraction of 2,4-TDA solved this problem prior to microbial cultivation, which enabled the recovery of this functional monomer while efficiently utilizing the remaining monomer mixture by a mixed culture. As proof-of-principle, mono-rhamnolipids were produced from an extracted mock PU hydrolysate, demonstrating its applicability as a carbon source for a biotechnological conversion. However, the process's intensification remains, such as the extractant's biocompatibility and recovery of 2,4-TDA from the extractant. In addition, this work presumes an efficient PU depolymerization process, which is also still in its infancy. If successful, new possibilities for the end-of-life upcycling of PUs would become available, sorely needed because this multimillion-ton polymer's current recycling rates are meager. Since the ALE strain of *P. capeferrum* TDA1 can utilize AA, 1,4-BDO, EG, and 2,4-TDA, this strain was also used to prove the concept of the consolidated utilization of PUs' monomers. Nevertheless, it was shown from the toxicity test that 2,4-TDA affected the decrease of growth compared to the control without 2,4-TDA. From this study, TDA1 strains can be used not only for the utilization of PU and PET monomers but also considered as the cell factory in the production of various valuable products, such as rhamnolipids, phenazine, and PHA. Since *P. capeferrum* TDA1 is also tolerant towards high temperature, adaptable to a toxic compound, and could grow on the broad spectrum of carbon sources, this strain could become a promising alternative in the industrial or environmental application of plastic waste upcycling using fermenter or bioreactor and coupled with some advanced technology in fermentation realm.

References

A. Glaser, J. (2019) Biological Degradation of Polymers in the Environment. In *Plastics in the Environment*. IntechOpen, pp. 1–22.

Ackermann, Y.S., Li, W.-J., Op de Hipt, L., Niehoff, P.-J., Casey, W., Polen, T., et al. (2021) Engineering adipic acid metabolism in Pseudomonas putida. *Metab Eng* **67**: 29–40.

Agler, M.T., Wrenn, B.A., Zinder, S.H., and Angenent, L.T. (2011) Waste to bioproduct conversion with undefined mixed cultures: the carboxylate platform. *Trends Biotechnol* **29**: 70–78.

Álvarez-Barragán, J., Domínguez-Malfavón, L., Vargas-Suárez, M., González-Hernández, R., Aguilar-Osorio, G., and Loza-Tavera, H. (2016) Biodegradative Activities of Selected Environmental Fungi on a Polyester Polyurethane Varnish and Polyether Polyurethane Foams. *Appl Environ Microbiol* **82**: 5225–5235.

Andrady, A.L. and Neal, M.A. (2009) Applications and societal benefits of plastics. *Philos Trans R Soc B Biol Sci* **364**: 1977–1984.

Antoniou, E., Fodelianakis, S., Korkakaki, E., and Kalogerakis, N. (2015) Biosurfactant production from marine hydrocarbon-degrading consortia and pure bacterial strains using crude oil as carbon source. *Front Microbiol* **06**: 1–14.

Askitosari, T.D., Berger, C., Tiso, T., Harnisch, F., Blank, L.M., and Rosenbaum, M.A. (2020) Coupling an Electroactive Pseudomonas putida KT2440 with Bioelectrochemical Rhamnolipid Production. *Microorganisms* **8**: 1959.

Askitosari, T.D., Boto, S.T., Blank, L.M., and Rosenbaum, M.A. (2019) Boosting Heterologous Phenazine Production in Pseudomonas putida KT2440 Through the Exploration of the Natural Sequence Space. *Front Microbiol* **10**: 1990.

Auman, H.J., Ludwig, J.P., Giesy, J.P., and Colborn, T. (1997) Plastic ingestion by Laysan Albatross chicks on Sand Island , Midway Atoll , in 1994 and 1995 . *Albatross Biol Conserv* 239–244.

Bai, B., Ren, J., Bai, F., and Hao, L. (2020) Selection and validation of reference genes for gene expression studies in Pseudomonas brassicacearum GS20 using real-time quantitative reverse transcription PCR. *PLoS One* **15**: e0227927.

Barnes, D.K.A., Galgani, F., Thompson, R.C., and Barlaz, M. (2009) Accumulation and fragmentation of plastic debris in global environments. *Philos Trans R Soc B Biol Sci* **364**: 1985–1998.

Bator, I., Karmainski, T., Tiso, T., and Blank, L.M. (2020) Killing two birds with one stone – Strain engineering facilitates the development of a unique rhamnolipid production process. *Front Bioeng Biotechnol* **8**: 899.

Beckers, V., Poblete-Castro, I., Tomasch, J., and Wittmann, C. (2016) Integrated analysis of gene expression and metabolic fluxes in PHA-producing Pseudomonas putida grown on glycerol. *Microb Cell Fact* **15**: 73.

Bednarz, A., Spieß, A.C., and Pfennig, A. (2017) Reactive and physical extraction of bio-based diamines from fermentation media. *J Chem Technol Biotechnol* **92**: 1817–1824.

Behrens, B., Baune, M., Jungkeit, J., Tiso, T., Blank, L.M., and Hayen, H. (2016) High performance liquid chromatography-charged aerosol detection applying an inverse gradient for quantification of rhamnolipid biosurfactants. *J Chromatogr A* **1455**: 125–132.

Beuker, J., Barth, T., Steier, A., Wittgens, A., Rosenau, F., Henkel, M., and Hausmann, R. (2016) High titer heterologous rhamnolipid production. *AMB Express* **6**: 124.

Biswas, R.K., Habib, M.A., and Islam, M.N. (2000) Some physicochemical properties of (D2EHPA). 1. Distribution, dimerization, and acid dissociation constants of D2EHPA in a kerosene/0.10 kmol m-3 (Na+,H+)Cl- system and the extraction of Mn(II). *Ind Eng Chem Res* **39**: 155–160.

Blesken, C.C., Bator, I., Eberlein, C., Heipieper, H.J., Tiso, T., and Blank, L.M. (2020) Genetic Cell-Surface Modification for Optimized Foam Fractionation. *Front Bioeng Biotechnol* **8**: 572892.

Blesken, C.C., Strümpfler, T., Tiso, T., and Blank, L.M. (2020) Uncoupling Foam Fractionation and Foam Adsorption for Enhanced Biosurfactant Synthesis and Recovery. *Microorganisms* **8**: 2029.

BMU (2018) Waste Management in Germany: Facts, data, diagrams.

Bongartz, P., Bator, I., Baitalow, K., Keller, R., Tiso, T., Blank, L.M., and Wessling, M. (2021) A scalable bubble-free membrane aerator for biosurfactant production. *Biotechnol Bioeng* **118**: 3545–3558.

Borrero-de Acuña, J.M., Aravena-Carrasco, C., Gutierrez-Urrutia, I., Duchens, D., and Poblete-Castro, I. (2019) Enhanced synthesis of medium-chain-length poly(3-hydroxyalkanoates) by inactivating the tricarboxylate transport system of Pseudomonas putida KT2440 and process development using waste vegetable oil. *Process Biochem* **77**: 23–30.

Borrero, N. V., Bai, F., Perez, C., Duong, B.Q., Rocca, J.R., Jin, S., and Huigens III, R.W. (2014)

Phenazine antibiotic inspired discovery of potent bromophenazine antibacterial agents against Staphylococcus aureus and Staphylococcus epidermidis. *Org Biomol Chem* **12**: 881–886.

Brooks, A.L., Wang, S., and Jambeck, J.R. (2018) The Chinese import ban and its impact on global plastic waste trade. *Sci Adv* **4**: eaat0131.

Cárdenas Espinosa, M.J., Blanco, A.C., Schmidgall, T., Atanasoff-Kardjalieff, A.K., Kappelmeyer, U., Tischler, D., et al. (2020) Toward biorecycling: isolation of a soil bacterium that grows on a polyurethane oligomer and monomer. *Front Microbiol* **11**: 1–10.

Carta, D., Cao, G., and D'Angeli, C. (2003) Chemical recycling of poly(ethylene terephthalate) (PET) by hydrolysis and glycolysis. *Environ Sci Pollut Res* **10**: 390–394.

Choi, K.H., Gaynor, J.B., White, K.G., Lopez, C., Bosio, C.M., Karkhoff-Schweizer, R.A.R., and Schweizer, H.P. (2005) A Tn7-based broad-range bacterial cloning and expression system. *Nat Methods* **2**: 443–448.

Choi, Kumar, A., and Schweizer, H.P. (2006) A 10-min method for preparation of highly electrocompetent Pseudomonas aeruginosa cells: Application for DNA fragment transfer between chromosomes and plasmid transformation. *J Microbiol Methods* **64**: 391–397.

Choi, Y.J., Bourque, D., Morel, L., Groleau, D., and Míguez, C.B. (2006) Multicopy integration and expression of heterologous genes in methylobacterium extorquens ATCC 55366. *Appl Environ Microbiol* **72**: 753–759.

Ciesielski, S., Możejko, J., and Przybyłek, G. (2010) The influence of nitrogen limitation on mcl-PHA synthesis by two newly isolated strains of Pseudomonas sp. *J Ind Microbiol Biotechnol* 511–520.

Clean Water Action (2021) The problem of marine plastic pollution.

Cregut, M., Bedas, M., Durand, M.J., and Thouand, G. (2013) New insights into polyurethane biodegradation and realistic prospects for the development of a sustainable waste recycling process. *Biotechnol Adv* **31**: 1634–1647.

Crépin, S., Harel, S., and Dozois, C.M. (2012) Chromosomal complementation using Tn7 transposon vectors in enterobacteriaceae. *Appl Environ Microbiol* **78**: 6001–6008.

DePristo, M.A., Banks, E., Poplin, R., Garimella, K. V, Maguire, J.R., Hartl, C., et al. (2011) A framework for variation discovery and genotyping using next-generation DNA sequencing data. *Nat Genet* **43**: 491–498.

Dikma Technologies Inc (2018) ProElut^{TM} LLE+ (Liquid-Liquid Extraction).

Dragosits, M. and Mattanovich, D. (2013) Adaptive laboratory evolution – principles and applications for biotechnology. *Microb Cell Fact* **12**: 64.

Dussud, C. and Ghiglione, J.F. (2014) Bacterial degradation of synthetic plastics. *Mar litter Mediterr Black Seas* **46**: 43–48.

Echa (2020) Paraffin oils.

European Commission (2019) A circular economy for plastics – Insights from research and innovation to inform policy and funding decisions.

European Union Risk Assessment Report (2008) 4-Methyl-m-phenylenediamine (toluene-2,4-diamine) risk assessment.

Franden, M.A., Jayakody, L.N., Li, W.J., Wagner, N.J., Cleveland, N.S., Michener, W.E., et al. (2018) Engineering Pseudomonas putida KT2440 for efficient ethylene glycol utilization. *Metab Eng* **48**: 197–207.

Freedman DL, Shanley RS, S.R. (1996) Aerobic biodegradation of 2,4_dinitrotoluene, aminonitrotoluene isomers, and 2,4-diaminotoluene. *J Hazard Mater* **49**: 1–14.

Gama, N., Ferreira, A., and Barros-Timmons, A. (2018) Polyurethane foams: past, present, and future. *Materials (Basel)* **11**: 1–35.

Garrido, M.A. and Font, R. (2015) Pyrolysis and combustion study of flexible polyurethane foam. *J Anal Appl Pyrolysis* **113**: 202–215.

Gawande, A., Zamre, G.S., Renge, V.C., Bharsakale, G.R., and Tayde, S. (2012) Utilization of waste plastic in asphalting of roads. *Sci Revs Chem Commun* **2**: 147–157.

Geyer, R., Jambeck, J.R., and Law, K.L. (2017) Production, use, and fate of all plastics ever made. *Sci Adv* **3**: 25–29.

Gomes, A.É.I., Stuchi, L.P., Siqueira, N.M.G., Henrique, J.B., Vicentini, R., Ribeiro, M.L., et al. (2018) Selection and validation of reference genes for gene expression studies in Klebsiella pneumoniae using Reverse Transcription Quantitative real-time PCR. *Sci Rep* **8**: 9001.

Grigore, M.E. (2017) Methods of recycling, properties and applications of recycled thermoplastic polymers. *Recycling* **2**:.

Guzik, M.W., Nitkiewicz, T., Wojnarowska, M., Sołtysik, M., Kenny, S.T., Babu, R.P., et al. (2021) Robust process for high yield conversion of non-degradable polyethylene to a biodegradable plastic using a chemo-biotechnological approach. *Waste Manag* **135**: 60–69.

Hamad, M.N.F., Marrez, D.A., and El-Sherbieny, S.M.R. (2020) Toxicity evaluation and antimicrobial activity of purified pyocyanin from Pseudomonas aeruginosa. *Biointerface Res Appl Chem* **10**: 6974–6990.

Hara, H., Eltis, L.D., Davies, J.E., and Mohn, W.W. (2007) Transcriptomic analysis reveals a bifurcated terephthalate degradation pathway in Rhodococcus sp. strain RHA1. *J Bacteriol* **189**: 1641–1647.

Hartmans, S., Smits, J.P., Van der Werf, M.J., Volkering, F., and De Bont, J.A.M. (1989) Metabolism of styrene oxide and 2-phenylethanol in the styrene-degrading Xanthobacter strain 124X. *Appl Environ Microbiol* **55**: 2850–2855.

Heipieper, H.J., Neumann, G., Cornelissen, S., and Meinhardt, F. (2007) Solvent-tolerant bacteria for biotransformations in two-phase fermentation systems. *Appl Microbiol Biotechnol* **74**: 961–973.

Hellemans, J., Mortier, G., De Paepe, A., Speleman, F., and Vandesompele, J. (2007) qBase relative quantification framework and software for management and automated analysis of real-time quantitative PCR data. *Genome Biol* **8**: R19.

Hetemäki, L., Hanewinkel, M., Muys, B., Ollikainen, M., Palahí, M., and Trasobares, A. (2017) Leading the way to a European circular bioeconomy strategy.

Hosaka, M., Kamimura, N., Toribami, S., Mori, K., Kasai, D., Fukuda, M., and Masai, E. (2013) Novel tripartite aromatic acid transporter essential for terephthalate uptake in Comamonas sp. strain E6. *Appl Environ Microbiol* **79**: 6148–6155.

Hosseini, R., Kuepper, J., Koebbing, S., Blank, L.M., Wierckx, N., and de Winde, J.H. (2017) Regulation of solvent tolerance in Pseudomonas putida S12 mediated by mobile elements. *Microb Biotechnol* **10**: 1558–1568.

Huntsman (2010) Elastomers: A guide to thermoplastic polyurethanes (TPU). 1–25.

Ignatyev, I.A., Thielemans, W., and Vander Beke, B. (2014) Recycling of polymers: A review. *ChemSusChem* **7**: 1579–1593.

ISO 20776-1:2019(E) (2019) Susceptibility testing of infectious agents and evaluation of performance of antimicrobial susceptibility test devices — Part 1: Broth micro-dilution reference method for testing the in vitro activity of antimicrobial agents against rapidly growing aerobi.

Jain, A. (2017) Waste Management in ASEAN Countries.

Jambeck, J.R., Geyer, R., Wilcox, C., Siegler, T.R., Perryman, M., Andrady, A., et al. (2015) Plastic waste inputs from land into the ocean. *Science (80-)* **347**: 768–771.

Jiang, Q., Li, Z., Cui, Z., Wei, R., Nie, K., Xu, H., and Liu, L. (2021) Quantum Mechanical Investigation of the Oxidative Cleavage of the C–C Backbone Bonds in Polyethylene Model Molecules. *Polymers (Basel)* **13**: 2730.

Kamimura, N., Aoyama, T., Yoshida, R., Takahashi, K., Kasai, D., Abe, T., et al. (2010) Characterization of the protocatechuate 4,5-cleavage pathway operon in comamonas sp. strain e6 and discovery of a novel pathway gene. *Appl Environ Microbiol* **76**: 8093–8101.

Kasai, D., Kitajima, M., Fukuda, M., and Masai, E. (2010) Transcriptional regulation of the terephthalate catabolism operon in comamonas sp. strain E6. *Appl Environ Microbiol* **76**: 6047–6055.

Kenny, S.T., Runic, J.N., Kaminsky, W., Woods, T., Babu, R.P., Keely, C.M., et al. (2008) Up-cycling of PET (Polyethylene Terephthalate) to the biodegradable plastic PHA (Polyhydroxyalkanoate). *Environ Sci Technol* **42**: 7696–7701.

Kenny, S.T., Runic, J.N., Kaminsky, W., Woods, T., Babu, R.P., and O'Connor, K.E. (2012) Development of a bioprocess to convert PET derived terephthalic acid and biodiesel derived glycerol to medium chain length polyhydroxyalkanoate. *Appl Microbiol Biotechnol* **95**: 623–633.

Khademi, S.M.H., Sazinas, P., and Jelsbak, L. (2019) Within-host adaptation mediated by intergenic evolution in Pseudomonas aeruginosa. *Genome Biol Evol* **11**: 1385–1397.

Khalid, A., Arshad, M., and Crowley, D.E. (2009) Biodegradation potential of pure and mixed bacterial cultures for removal of 4-nitroaniline from textile dye wastewater. *Water Res* **43**: 1110–1116.

Khan, S., Nadir, S., Shah, Z.U., Shah, A.A., Karunarathna, S.C., Xu, J., et al. (2017) Biodegradation of polyester polyurethane by Aspergillus tubingensis. *Environ Pollut* **225**: 469–480.

Kleerebezem, R. and van Loosdrecht, M.C. (2007) Mixed culture biotechnology for bioenergy production. *Curr Opin Biotechnol* **18**: 207–212.

Krishnaiah, M., de Almeida, N.R., Udumula, V., Song, Z., Chhonker, Y.S., Abdelmoaty, M.M., et al. (2018) Synthesis, biological evaluation, and metabolic stability of phenazine derivatives as antibacterial agents. *Eur J Med Chem* **143**: 936–947.

De Las Heras, A., Carreño, C.A., and De Lorenzo, V. (2008) Stable implantation of orthogonal sensor circuits in Gram-negative bacteria for environmental release. *Environ Microbiol*

10: 3305–3316.

Law, K.L., Starr, N., Siegler, T.R., Jambeck, J.R., Mallos, N.J., and Leonard, G.H. (2020) The United States' contribution of plastic waste to land and ocean. *Sci Adv* **6**: eabd0288.

Lebreton, L.C.M., van der Zwet, J., Damsteeg, J., Slat, B., Andrady, A., and Reisser, J. (2017) River plastic emissions to the world's oceans. *Nat Commun* **8**: 15611.

Li, X., Chen, Y., & Nielsen, J. (2019) Harnessing xylose pathways for biofuels production. ,. *Curr Opin Biotechnol* **57**: 56–65.

Li, W.-J. (2020) Plastic monomer degradation Engineering Pseudomonas putida KT2440 for plastic monomer utilization.

Li, W.J., Jayakody, L.N., Franden, M.A., Wehrmann, M., Daun, T., Hauer, B., et al. (2019) Laboratory evolution reveals the metabolic and regulatory basis of ethylene glycol metabolism by Pseudomonas putida KT2440. *Environ Microbiol* **21**: 3669–3682.

Li, W.J., Narancic, T., Kenny, S.T., Niehoff, P.J., O'Connor, K., Blank, L.M., and Wierckx, N. (2020) Unraveling 1,4-butanediol metabolism in Pseudomonas putida KT2440. *Front Microbiol* **11**: 1–15.

Liu, P., Zhu, X., Tan, Z., Zhang, X., and Ma, Y. (2015) Construction of Escherichia coli cell Factories for production of organic acids and alcohols., pp. 107–140.

Liu, Y.-J., Li, B., Feng, Y., and Cui, Q. (2020) Consolidated bio-saccharification: Leading lignocellulose bioconversion into the real world. *Biotechnol Adv* **40**: 107535.

Livak, K.J. and Schmittgen, T.D. (2001) Analysis of relative gene expression data using real-time quantitative PCR and the 2–ΔΔCT method. *Methods* **25**: 402–408.

Maal, A., Ali, N., Ramli, I., and Universit, H. (2017) Experimental study the performance of asphalt concrete which using plastics powder filler fill in submersed water conditions. *Int J Civ Eng Technol* **8**: 686–696.

Magnin, A., Entzmann, L., Bazin, A., Pollet, E., and Avérous, L. (2021) Green recycling process for polyurethane foams by a chem-biotech approach. *ChemSusChem* **14**: 4234–4241.

Magnin, A., Entzmann, L., Pollet, E., and Avérous, L. (2021) Breakthrough in polyurethane bio-recycling: An efficient laccase-mediated system for the degradation of different types of polyurethanes. *Waste Manag* **132**: 23–30.

Magnin, A., Hoornaert, L., Pollet, E., Laurichesse, S., Phalip, V., and Avérous, L. (2019) Isolation and characterization of different promising fungi for biological waste management of polyurethanes. *Microb Biotechnol* **12**: 544–555.

Magnin, A., Pollet, E., Perrin, R., Ullmann, C., Persillon, C., Phalip, V., and Avérous, L. (2019) Enzymatic recycling of thermoplastic polyurethanes: Synergistic effect of an esterase and an amidase and recovery of building blocks. *Waste Manag* **85**: 141–150.

Magnin, A., Pollet, E., Phalip, V., and Avérous, L. (2020) Evaluation of biological degradation of polyurethanes. *Biotechnol Adv* **39**: 107457.

Mattioda, G. and Christidis, Y. (2000) Glyoxylic Acid. In *Ullmann's Encyclopedia of Industrial Chemistry*. Weinheim, Germany: Wiley-VCH Verlag GmbH & Co. KGaA.

McIlgorm, A., Campbell, H.F., and Rule, M.J. (2009) Understanding the economic benefits and costs of controlling marine debris in the APEC region.

McKenna, A., Hanna, M., Banks, E., Sivachenko, A., Cibulskis, K., Kernytsky, A., et al. (2010) The Genome Analysis Toolkit: A MapReduce framework for analyzing next-generation DNA sequencing data. *Genome Res* **20**: 1297–1303.

Meier, K. (2020) Optimization of Pseudomonas sp. TDA1 for co-utilization of PET monomers.

Modak, P., Pariatamby, A., and Seadon, J. (2017) Asia waste management outlook.

Mohamed, E.T., Werner, A.Z., Salvachúa, D., Singer, C.A., Szostkiewicz, K., Rafael Jiménez-Díaz, M., et al. (2020) Adaptive laboratory evolution of Pseudomonas putida KT2440 improves p-coumaric and ferulic acid catabolism and tolerance. *Metab Eng Commun* **11**: 1–14.

Molitor, B., Richter, H., Martin, M.E., Jensen, R.O., Juminaga, A., Mihalcea, C., and Angenent, L.T. (2016) Carbon recovery by fermentation of CO-rich off gases – Turning steel mills into biorefineries. *Bioresour Technol* **215**: 386–396.

Mückschel, B., Simon, O., Klebensberger, J., Graf, N., Rosche, B., Altenbuchner, J., et al. (2012) Ethylene glycol metabolism by Pseudomonas putida. *Appl Environ Microbiol* **78**: 8531–8539.

Müller, M.M., Hörmann, B., Syldatk, C., and Hausmann, R. (2010) Pseudomonas aeruginosa PAO1 as a model for rhamnolipid production in bioreactor systems. *Appl Microbiol Biotechnol* **87**: 167–174.

Narancic, T., Salvador, M., Hughes, G.M., Beagan, N., Abdulmutalib, U., Kenny, S.T., et al. (2021) Genome analysis of the metabolically versatile Pseudomonas umsongensis GO16: the genetic basis for PET monomer upcycling into polyhydroxyalkanoates. *Microb Biotechnol*.

Narancic, T., Verstichel, S., Reddy Chaganti, S., Morales-Gamez, L., Kenny, S.T., De Wilde, B., et al. (2018) Biodegradable plastic blends create new possibilities for end-of-life management of plastics but they are not a panacea for plastic pollution. *Environ Sci Technol* **52**: 10441–10452.

Nies, S.C., Alter, T.B., Nölting, S., Thiery, S., Phan, A.N.T., Drummen, N., et al. (2020) High titer methyl ketone production with tailored Pseudomonas taiwanensis VLB120. *Metab Eng* **62**: 84–94.

Nikje, M.M.A., Garmarudi, A.B., and Idris, A.B. (2011) Polyurethane waste reduction and recycling: From bench to pilot scales. *Des Monomers Polym* **14**: 395–421.

Nikodinovic, J., Kenny, S.T., Babu, R.P., Woods, T., Blau, W.J., and O'Connor, K.E. (2008) The conversion of BTEX compounds by single and defined mixed cultures to medium-chain-length polyhydroxyalkanoate. *Appl Microbiol Biotechnol* **80**: 665–673.

Ochsner, U.A., Reiser, J., Fiechter, A., and Witholt, B. (1995) Production of Pseudomonas aeruginosa rhamnolipid biosurfactants in heterologous hosts. *Appl Environ Microbiol* **61**: 3503–3506.

Onbasli, D. and Aslim, B. (2009) Biosurfactant production in sugar beet molasses by some Pseudomonas spp. *J Env Biol* **30**: 161–163.

Osman, M., Satti, S.M., Luqman, A., Hasan, F., Shah, Z., and Shah, A.A. (2018) Degradation of polyester polyurethane by Aspergillus sp. strain S45 isolated from soil. *J Polym Environ* **26**: 301–310.

Ostiategui-Francia, P., Usategui-Martín, A., and Liria-Loza, A. (2016) Microplastics presence in Sea Turtles. In *Fate and Impact of Microplastics in Marine Ecosystems*. Lanzarote: Elsevier, pp. 34–35.

Parke, D., Garcia, M.A., and Ornston, L.N. (2001) Cloning and genetic characterization of dca genes required for β-oxidation of straight-chain dicarboxylic acids in Acinetobacter sp. strain ADP1. *Appl Environ Microbiol* **67**: 4817–4827.

Pfaffl, M.W. (2001) A new mathematical model for relative quantification in real-time RT-PCR. *Nucleic Acids Res* **29**: 45e – 45.

Plastics Europe (2019) Plastics – the Facts 2019. An analysis of European plastics production, demand and waste data.

Plastics Europe (2020) Plastics – the Facts 2020. An analysis of European plastics production, demand and waste data.

PlasticsEurope (2012) Toluene Diisocyanate (TDI) & Methylenediphenyl Diisocyanate (MDI).

Poblete-Castro, I., Becker, J., Dohnt, K., dos Santos, V.M., and Wittmann, C. (2012) Industrial biotechnology of Pseudomonas putida and related species. *Appl Microbiol Biotechnol* **93**: 2279–2290.

Poblete-Castro, I., Escapa, I.F., Jäger, C., Puchalka, J., Chi Lam, C., Schomburg, D., et al. (2012) The metabolic response of P. putida KT2442 producing high levels of polyhydroxyalkanoate under single- and multiple-nutrient-limited growth: Highlights from a multi-level omics approach. *Microb Cell Fact* **11**: 34.

Polystyreneloop (2021) Technology: One of the key options for dealing with the growing volumes of end of life construction PS foam waste streams in the decades to come is the closed-loop recycling of PS foam waste.

Ramakers, C., Ruijter, J.M., Deprez, R.H.L., and Moorman, A.F.. (2003) Assumption-free analysis of quantitative real-time polymerase chain reaction (PCR) data. *Neurosci Lett* **339**: 62–66.

Ritchie, H. and Roser, M. (2018) Plastic Pollution. *Our World Data*.

Ruijter, J.M., Ramakers, C., Hoogaars, W.M.H., Karlen, Y., Bakker, O., van den Hoff, M.J.B., and Moorman, A.F.M. (2009) Amplification efficiency: linking baseline and bias in the analysis of quantitative PCR data. *Nucleic Acids Res* **37**: e45–e45.

Salvador, M., Abdulmutalib, U., Gonzalez, J., Kim, J., Smith, A.A., Faulon, J.-L., et al. (2019) Microbial genes for a circular and sustainable bio-PET economy. *Genes (Basel)* **10**: 373.

Sánchez-Clemente, R., Igeño, M.I., Población, A.G., Guijo, M.I., Merchán, F., and Blasco, R. (2018) Study of pH changes in media during bacterial growth of several environmental strains. *Proceedings* **2**: 1297.

Sasoh, M., Masai, E., Ishibashi, S., Hara, H., Kamimura, N., Miyauchi, K., and Fukuda, M. (2006) Characterization of the terephthalate degradation genes of Comamonas sp. strain E6. *Appl Environ Microbiol* **72**: 1825–1832.

Schlembach, I., Naturstoff-forschung, L., Blank, L.M., and Wierckx, N. (2020) Consolidated bioprocessing of cellulose to itaconic acid by a co-culture of Trichoderma reesei and Ustilago maydis. 1–21.

Schmidt, J., Wei, R., Oeser, T., Dedavid e Silva, L., Breite, D., Schulze, A., and Zimmermann, W. (2017) Degradation of Polyester Polyurethane by Bacterial Polyester Hydrolases. *Polymers (Basel)* **9**: 1–10.

Schmitz, S., Nies, S., Wierckx, N., Blank, L.M., and Rosenbaum, M.A. (2015) Engineering mediator-based electroactivity in the obligate aerobic bacterium Pseudomonas putida

KT2440. *Front Microbiol* **6**: 1–13.

Shakorfow, A.M. (2016) Biomass. Incineration, pyrolysis, combustion and gasification. *Int J Sci Res* **5**: 13–25.

Sheel, A. and Pant, D. (2018) Chemical depolymerization of polyurethane foams via glycolysis and hydrolysis, Elsevier Inc.

Skoog, E., Shin, J.H., Saez-Jimenez, V., Mapelli, V., and Olsson, L. (2018) Biobased adipic acid – The challenge of developing the production host. *Biotechnol Adv* **36**: 2248–2263.

Skorupa, B. (2020) Optimization of adipic acid utilization in Pseudomonas sp. TDA1.

Song, Y., Wei, J., Ma, Y., Zeng, P., and Kong, M. (2015) Removal and recovery of amantadine from water by liquid–liquid extraction. *Environ Earth Sci* **73**: 4931–4938.

Songwon (2017) Polyester Diols. 1-8.

Sonnenschein, M.F. (2014) Polyurethanes, First. Grossman R.F., N.D. (ed) Hoboken, NJ: John Wiley & Sons, Inc.

Spiekermann, P., Rehm, B.H.A., Kalscheuer, R., Baumeister, D., and Steinbüchel, A. (1999) A sensitive, viable-colony staining method using Nile red for direct screening of bacteria that accumulate polyhydroxyalkanoic acids and other lipid storage compounds. *Arch Microbiol* **171**: 73–80.

Sulaiman, S., Yamato, S., Kanaya, E., Kim, J.-J., Koga, Y., Takano, K., and Kanaya, S. (2012) Isolation of a novel cutinase homolog with polyethylene terephthalate-degrading activity from leaf-branch compost by using a metagenomic approach. *Appl Environ Microbiol* **78**: 1556–1562.

Suvorova, I.A., Korostelev, Y.D., and Gelfand, M.S. (2015) GntR family of bacterial transcription factors and their DNA binding motifs: Structure, positioning and co-evolution. *PLoS One* **10**: e0132618.

Szycher, M. (2012) Structure–property relations in polyurethanes. In *Szycher's Handbook of Polyurethanes, Second Edition*. CRC Press, pp. 37–86.

Taylor-Smith, K. (2018) How plastic pollution is affecting the ocean wildlife.

The Center for Biological Diversity (2018) Ocean plastics pollution.

Thorvaldsdottir, H., Robinson, J.T., and Mesirov, J.P. (2013) Integrative Genomics Viewer (IGV): high-performance genomics data visualization and exploration. *Brief Bioinform* **14**: 178–192.

Tiso, T., Narancic, T., Wei, R., Pollet, E., Beagan, N., Schröder, K., et al. (2021) Towards bio-upcycling of polyethylene terephthalate. *Metab Eng* **66**: 167–178.

Tiso, T., Sabelhaus, P., Behrens, B., Wittgens, A., Rosenau, F., Hayen, H., and Blank, L.M. (2016) Creating metabolic demand as an engineering strategy in Pseudomonas putida – Rhamnolipid synthesis as an example. *Metab Eng Commun* **3**: 234–244.

Tiso, T., Zauter, R., Tulke, H., Leuchtle, B., Li, W.J., Behrens, B., et al. (2017) Designer rhamnolipids by reduction of congener diversity: Production and characterization. *Microb Cell Fact* **16**: 1–14.

Toprak, E., Veres, A., Michel, J., Chait, R., Hartl, D.L., and Kishony, R. (2012) Evolutionary paths to antibiotic resistance under dynamically sustained drug selection. *Nat Genet* **44**: 101–105.

Tournier, V., Topham, C.M., Gilles, A., David, B., Folgoas, C., Moya-Leclair, E., et al. (2020) An engineered PET depolymerase to break down and recycle plastic bottles. *Nature* **580**: 216–219.

Udvardi, M.K., Czechowski, T., and Scheible, W.-R. (2008) Eleven golden rules of quantitative RT-PCR. *Plant Cell* **20**: 1736–1737.

Umweltbundesamt (2008) The role of waste incineration in Germany.

Umweltbundesamt (2014) Waste regulations.

Utomo, R.N.C., Heipieper, H.J., Eberlein, C., and Blank, L.M. (2021) New species Pseudomonas capeferrum TDA1 as a plastic monomer utilizer and a PHA native producer. *IOP Conf Ser Earth Environ Sci* **894**: 012024.

Utomo, R.N.C., Li, W.-J., Tiso, T., Eberlein, C., Doeker, M., Heipieper, H.J., et al. (2020) Defined microbial mixed culture for utilization of polyurethane monomers. *ACS Sustain Chem Eng* **8**: 17466–17474.

Vague, M., Chan, G., Roberts, C., Swartz, N.A., and Mellies, J.L. (2019) Pseudomonas isolates degrade and form biofilms on polyethylene terephthalate (PET) plastic. *Prepr biorxiv.*

Valentine-King, M.A., Cisneros, K., James, M.O., Huigens, R.W., and Brown, M.B. (2020) Efficacy data of halogenated phenazine and quinoline agents and an NH125 analogue to veterinary mycoplasmas. *BMC Vet Res* **16**: 107.

Vandesompele, J., De Preter, K., Pattyn, F., Poppe, B., Van Roy, N., De Paepe, A., and Speleman, F. (2002) Accurate normalization of real-time quantitative RT-PCR data by geometric averaging of multiple internal control genes. *Genome Biol* **3**: research0034.1.

Venkatachalam, S., G., S., V., J., R., P., Rao, K., and K., A. (2012) Degradation and Recyclability of Poly (Ethylene Terephthalate). *Polyester* 75–97.

Vinuselvi, P., Kim, M.K., Lee, S.K., and Ghim, C.M. (2012) Rewiring carbon catabolite repression for microbial cell factory. *BMB Rep* **45**: 59–70.

Vinylplus (2014) PVC recycling technologies. *Vinylplus* 1–48.

Wang, G., Huang, D., Ji, J., Völker, C., and Wurm, F.R. (2021) Seawater-degradable polymers—Fighting the marine plastic pollution. *Adv Sci* **8**: 2001121.

Wang, X., Xia, K., Yang, X., and Tang, C. (2019) Growth strategy of microbes on mixed carbon sources. *Nat Commun* **10**: 1–7.

Wang, Y.Z., Zhou, Y., and Zylstra, G.J. (1995) Molecular analysis of isophthalate and terephthalate degradation by comamonas testosteroni YZW-D. *Environ Health Perspect* **103**: 9–12.

Watson, T.G. (1972) The present status and future prospects of the turbidostat. *J Appl Chem Biotechnol* **22**: 229–243.

Webb, H., Arnott, J., Crawford, R., and Ivanova, E. (2012) Plastic degradation and its environmental implications with special reference to poly(ethylene terephthalate). *Polymers (Basel)* **5**: 1–18.

Wei, R., Breite, D., Song, C., Gräsing, D., Ploss, T., Hille, P., et al. (2019) Biocatalytic degradation efficiency of postconsumer polyethylene terephthalate packaging determined by their polymer microstructures. *Adv Sci* **6**:.

Wei, R. and Zimmermann, W. (2017a) Biocatalysis as a green route for recycling the recalcitrant plastic polyethylene terephthalate. *Microb Biotechnol* **10**: 1302–1307.

Wei, R. and Zimmermann, W. (2017b) Microbial enzymes for the recycling of recalcitrant petroleum-based plastics: how far are we? *Microb Biotechnol* **10**: 1308–1322.

Wierckx, N., Narancic, T., Eberlein, C., Wei, R., Drzyzga, O., Magnin, A., et al. (2018) Plastic biodegradation: Challenges and opportunities. In *Consequences of Microbial Interactions with Hydrocarbons, Oils, and Lipids: Biodegradation and Bioremediation*. Cham: Springer International Publishing, pp. 1–29.

Wierckx, N., Prieto, M.A., Pomposiello, P., de Lorenzo, V., O'Connor, K., and Blank, L.M. (2015) Plastic waste as a novel substrate for industrial biotechnology. *Microb Biotechnol* **8**: 900–903.

Wilkes, R.A. and Aristilde, L. (2017) Degradation and metabolism of synthetic plastics and associated products by Pseudomonas sp.: capabilities and challenges. *J Appl Microbiol*

123: 582–593.

World Energy Council (2016) World energy resources. Waste to energy. 2016.

Worm, B., Lotze, H.K., Jubinville, I., Wilcox, C., and Jambeck, J. (2017) Plastic as a persistent marine pollutant. *Annu Rev Environ Resour* **42**: 1–26.

Wynands, B., Otto, M., Runge, N., Preckel, S., Polen, T., Blank, L.M., and Wierckx, N. (2019) Streamlined Pseudomonas taiwanensis VLB120 Chassis Strains with Improved Bioprocess Features. *ACS Synth Biol* **8**: 2036–2050.

Xanthos, M. and Patel, S.H. (1998) Solvolysis. In *Frontiers in the Science and Technology of Polymer Recycling*. Dordrecht: Springer Netherlands, pp. 425–436.

Yang, Y.T. (2018) Origin and evolution of biodiversity, Pontarotti, P. (ed) Cham: Springer International Publishing.

Yoshida, S., Hiraga, K., Takehana, T., Taniguchi, I., Yamaji, H., Maeda, Y., et al. (2016) A bacterium that degrades and assimilates poly(ethylene terephthalate). *Science (80-)* **351**: 1196–1199.

Yu, J.-L., Xia, X.-X., Zhong, J.-J., and Qian, Z.-G. (2014) Direct biosynthesis of adipic acid from a synthetic pathway in recombinant Escherichia coli. *Biotechnol Bioeng* **111**: 2580–2586.

Zhang, X., Fevre, M., Jones, G.O., and Waymouth, R.M. (2018) Catalysis as an enabling science for sustainable polymers. *Chem Rev* **118**: 839–885.

Zhu, K. and Rock, C.O. (2008) RhlA converts β-hydroxyacyl-acyl carrier protein intermediates in fatty acid synthesis to the β-hydroxydecanoyl-β-hydroxydecanoate component of rhamnolipids in Pseudomonas aeruginosa. *J Bacteriol* **190**: 3147–3154.

Zou, W., Liu, L., and Chen, J. (2013) Structure, mechanism and regulation of an artificial microbial ecosystem for vitamin C production. *Crit Rev Microbiol* **39**: 247–255.

Appendix

Plasmid map

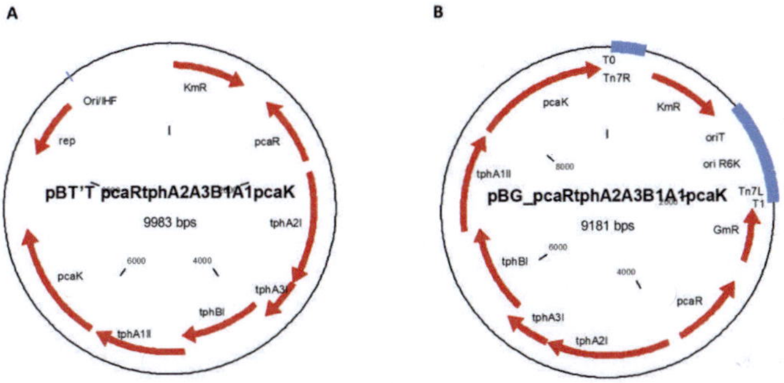

Figure S1 Plasmid construct harboring *tph* operon for terephthalic acid utilization. A: Construct for plasmid based was provided by Tanja Narancic. B: Construct for genome integration was provided by Sebastian Köbbing.

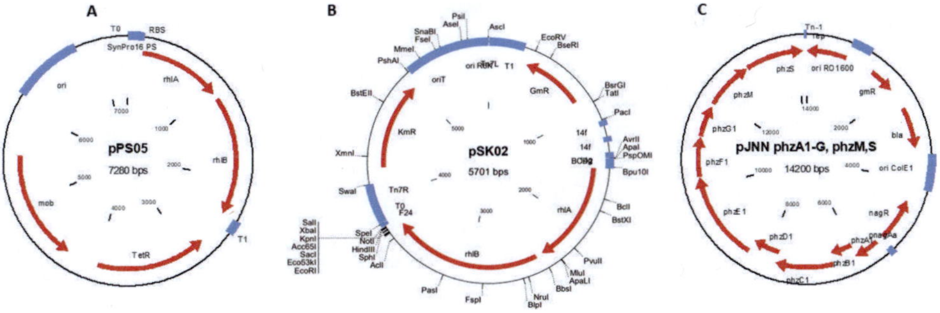

Figure S2 Plasmid construct for producing valuable products. Plasmid-based (A) and Tn7-based (B) constructs for mono-rhamnolipid production. The constructs were provided by Till Tiso. B: Construct for genome integration was provided by Simone Schmitz)

Table S1 Summary of identified mutations in laboratory *P. capeferrum* TDA1, ALE *P. capeferrum* TDA1, and ALE TDA1ge_*tph* F1.1

		Reference *P. capeferrum* TDA1 vs laboratory *P. capeferrum* TDA1	Laboratory *P. capeferrum* TDA1 vs	
			ALE TDA1	ALE TDA1ge_*tph* F1.1
Sum of mutations		45	39	68
Type of mutation	SNP	22	14	33
	InDel	23	25	35

Annotated essential genes in TDA1 genome

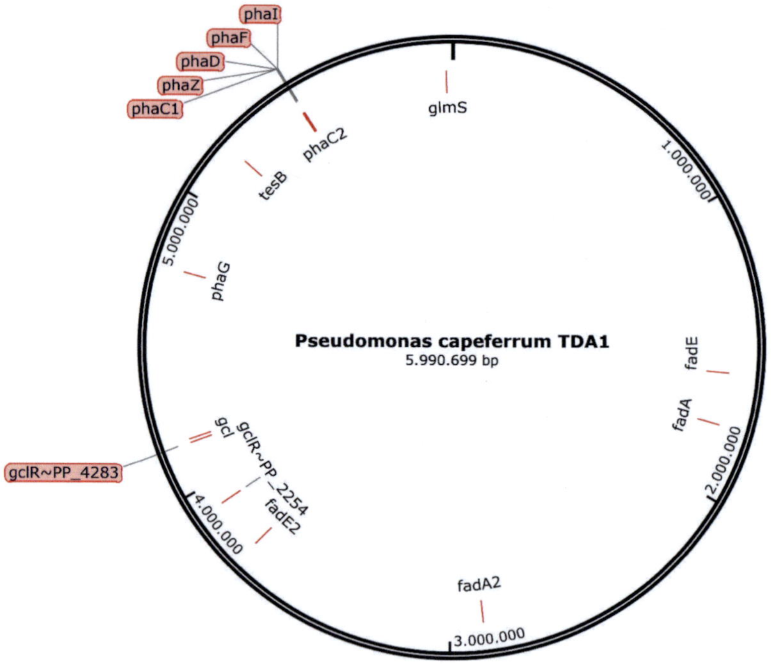

Figure S3 Complete genome of *P. capeferrum* TDA1 with annotated essential genes

Curriculum Vitae

Personal Information	
Name:	Romualdus Nugraha Catur Utomo
Born:	Malang, June 19th, 1988
Nationality:	Indonesian

Education	
2018-2022	Doctoral studies, Institute of Applied Microbiology, RWTH Aachen University, Aachen, Germany
2011-2013	Master of Engineering, Biology Engineering Department, INHA University, Incheon, Republic of Korea
2006-2010	Bachelor of Science, Biology Department, Faculty of Mathematics and Science, University of Brawijaya, Malang, Indonesia

Work Experience	
01/2018-01/2022	Doctoral Thesis, Institute of Applied Microbiology, RWTH Aachen University, Aachen, Germany
04/2019-present	Voluntary job as a reviewer, Journal of Tropical Biology, University of Brawijaya, Malang, Indonesia
01/2013-10/2017	Researcher, Research and Development Department, Cheil Jedang Indonesia Corp • Planning and conducting amino acid (lysine and tryptophan) fermentation • Responsible for scale-up implementation from lab scale to plant scale fermentation • Leading and coordinating foremans and operators • Person in charge of SOP and ISO documentation and implementation • Suggestion and Idea Program (SIP) team of Cheil Jedang Indonesia (CJI) • Internal auditor for ISO 50001 (energy management system)

02/2010-07/2011	Quality Control supervisor, Quality Control Department, Karunia Alam Segar Corp. (Wings Group) Quality Control Department • Leading and coordinating Quality Control coordinators and operators • Controlling the whole seasoning process: from raw materials to end products

Awards

24/08/2021	Best paper of conference, International Conference on Environmental and sustainability context, Semarang, Indonesia
01/2018-01/2022	Doctoral fellowship, Indonesia Endowment Fund for Education, Ministry of Finance of Republic of Indonesia
09/2011-08/2013	Jungseok International Scholarship for Master Program, INHA University, Incheon, Republic of Korea
09/2011-08/2013	Brain Korea 21 (BK 21) program for leading universities and students and Cheil Jedang funding for Master Program, INHA University, Incheon, Republic of Korea
2008-2010	Youth Leadership Scholarship, Youth Commision of Malang Diocese, Malang, Indonesia
2008-2010	Student Learning Help Scholarship, University of Brawijaya, Malang, Indonesia

Publications

2021	Romualdus Nugraha Catur Utomo, Hermann J. Heipieper, Christian Eberlein, and Lars M. Blank; "New species *Pseudomonas capeferrum* TDA1 as a plastic monomer utilizer and a PHA native producer." *IOP Conf. Series: Earth and Environmental Science* 2021, 894, 012024. DOI: 10.1088/1755-1315/894/1/012024.
2020	Romualdus Nugraha Catur Utomo, Wing-Jin Li, Till Tiso, Christian Eberlein, Moritz Doeker, Hermann J. Heipieper, Andreas Jupke, Nick Wierckx, and Lars M. Blank; "Defined microbial mixed Culture for utilization of polyurethane monomers" in ACS Sustainable Chemistry & Engineering in ACS (DOI:10.2021/acssuschemeng.0c06019).
2020	Romualdus Nugraha Catur Utomo, Sebastian Köbbing, Tanja Narancic, Kevin O'Connor, Nick Wierckx, Till Tiso, Lars M Blank; "Multi-purpose Pseudomonas B1 strain in downstream process of polyethylene terephthalate (PET) monomers for a circular bioeconomy" in Proceeding of 6th Joint Conference of DGHM &VAAM, 8-11 March 2020, Leipzig, Germany.
2017	Romualdus Nugraha Catur Utomo, Eun-Kwang Lee, and Hyun Shik Yun," Solid state fermentation of *Phanerochaete chrysosporium* for degradation and saccharification of lignocellulose" in Korean Society for Biotechnology and Bioengineering Journal (DOI: 10.7841/ksbbj.2017.32.2.96).
2013	Romualdus Nugraha Catur Utomo and Hyun Shik Yun, "Enzymatic activities of *Phanerochaete chrysosporium* in solid state fermentation (SSF) for lignin degradation and saccharification of spent coffee ground (SCG)" in Indonesian Scholars Journal Vol. 1 No.1.
2010	Suharjono, Sutrisno, Romualdus Nugraha Catur Utomo, "Biodegradasi linear alkylbenzene sulfonate (LAS) oleh konsorsium bakteri pembentuk biofilm di sedimen ekosistem air sungai" in Proceeding of National Seminar: Biodiversity and Biotechnology of

	Aquatic Resources, Jenderal Soedirman University, Purwokerto, Indonesia.

Oral Presentations

2021	New species *Pseudomonas capeferrum* TDA1 as a plastic monomer utilizer and a PHA native utilizer, 1st International Conference on Environmental and Sustainability Context (ICOESCO), Semarang, Indonesia.
2019	*Pseudomonas capeferrum* TDA1 as multi-purpose bacteria for PU's monomers utilization, 2nd Pseudomonas Grassroot Meeting, Leiden, Netherlands.

Poster Presentations

2020	*Pseudomonas capeferrum* TDA1 as multi-purpose bacteria for PET's monomers utilization, 6th Joint Conference of DGHM & VAAM, 8-11 March 2020, Leipzig, Germany.
2019	Evolved *Pseudomonas* sp. B1 strain as multi-purpose bacteria in polyurethanes (PU) monomer utilization, 2nd Pseudomonas Grassroot Meeting, Leiden, Netherlands.